AF819181

Section 20

Alternative Separation Processes*

Michael E. Prudich, Ph.D. *Professor of Chemical Engineering, Ohio University; Member, American Institute of Chemical Engineers, American Chemical Society, American Society for Engineering Education (Section Editor, Alternative Solid/Liquid Separations)*

Huanlin Chen, M.Sc. *Professor of Chemical and Biochemical Engineering, Zhejiang University (Selection of Biochemical Separation Processes—Affinity Membrane Chromatography)*

Tingyue Gu, Ph.D. *Associate Professor of Chemical Engineering, Ohio University (Selection of Biochemical Separation Processes)*

Ram B. Gupta, Ph.D. *Alumni (Chair) Professor of Chemical Engineering, Department of Chemical Engineering, Auburn University; Member, American Institute of Chemical Engineers, American Chemical Society (Supercritical Fluid Separation Processes)*

Keith P. Johnston, Ph.D., P.E. *M. C. (Bud) and Mary Beth Baird Endowed Chair and Professor of Chemical Engineering, University of Texas (Austin); Member, American Institute of Chemical Engineers, American Chemical Society, University of Texas Separations Research Program (Supercritical Fluid Separation Processes)*

Herb Lutz *Consulting Engineer, Millipore Corporation; Member, American Institute of Chemical Engineers, American Chemical Society (Membrane Separation Processes)*

Guanghui Ma, Ph.D. *Professor, State Key Laboratory of Biochemical Engineering, Institute of Process Engineering, CAS, Beijing, China (Selection of Biochemical Separation Processes—Gigaporous Chromatography Media)*

Zhiguo Su, Ph.D. *Professor and Director, State Key Laboratory of Biochemical Engineering, Institute of Process Engineering, CAS, Beijing, China (Selection of Biochemical Separation Processes—Protein Refolding, Expanded-Bed Chromatography)*

CRYSTALLIZATION FROM THE MELT

*The contributions of Dr. Joseph D. Henry (Alternative Solid/Liquid Separations), Dr. William Eykamp (Membrane Separation Processes), Dr. T. Alan Hatton (Selection of Biochemical Separation Processes), Dr. Robert Lemlich (Adsorptive-Bubble Separation Methods), Dr. Charles G. Moyers (Crystallization from the Melt), and Dr. Michael P. Thien (Selection of Biochemical Separation Processes), who were authors for the seventh edition, are acknowledged.

SUPERCRITICAL FLUID SEPARATION PROCESSES

ALTERNATIVE SOLID/LIQUID SEPARATIONS

MEMBRANE SEPARATION PROCESSES

SELECTION OF BIOCHEMICAL SEPARATION PROCESSES

CRYSTALLIZATION FROM THE MELT

GENERAL REFERENCES: Van't Land, *Industrial Crystallization of Melts*, Taylor & Francis, New York, 2004. Mullin, *Crystallization*, 4th ed., Butterworth-Heinemann, 2001. Myerson, *Handbook of Industrial Crystallization*, 2d ed., Butterworth-Heinemann, 2001. Pfann, *Zone Melting*, 2d ed., Wiley, New York, 1966. U.S. Patents 3,621,664 and 3,796,060. Zief and Wilcox, *Fractional Solidification*, Marcel Dekker, New York, 1967.

INTRODUCTION

Purification of a chemical species by solidification from a liquid mixture can be termed either **solution** crystallization or crystallization from the **melt.** The distinction between these two operations is somewhat subtle. The term **melt crystallization** has been defined as the separation of components of a binary mixture without addition of solvent, but this definition is somewhat restrictive. In **solution crystallization** a diluent solvent is added to the mixture; the solution is then directly or indirectly cooled, and/or solvent is evaporated to effect crystallization. The solid phase is formed and maintained somewhat below its pure-component freezing-point temperature. In melt crystallization no diluent solvent is added to the reaction mixture, and the solid phase is formed by cooling of the melt. Product is frequently maintained near or above its pure-component freezing point in the refining section of the apparatus.

A large number of techniques are available for carrying out crystallization from the melt. An abbreviated list includes partial freezing and solids recovery in cooling crystallizer-centrifuge systems, partial melting (e.g., sweating), staircase freezing, normal freezing, zone melting, and column crystallization. A description of all these methods is not within the scope of this discussion. Zief and Wilcox (op. cit.) and Myerson (op. cit.) describe many of these processes. Three of the more common methods—progressive freezing from a falling film, zone melting, and melt crystallization from the bulk—are discussed here to illustrate the techniques used for practicing crystallization from the melt.

High or ultrahigh product purity is obtained with many of the melt-purification processes. Table 20-1 compares the product quality and product form that are produced from several of these operations. Zone refining can produce very pure material when operated in a batch mode; however, other melt crystallization techniques also provide high purity and become attractive if continuous high-capacity processing is desired. Comparison of the features of melt crystallization and distillation are shown on Table 20-2.

A brief discussion of solid-liquid phase equilibrium is presented prior to discussing specific crystallization methods. Figures 20-1 and 20-2 illustrate the phase diagrams for binary solid-solution and eutectic systems, respectively. In the case of binary solid-solution systems, illustrated in Fig. 20-1, the liquid and solid phases contain equilibrium quantities of both components in a manner similar to vapor-liquid phase behavior. This type of behavior causes separation difficulties since multiple stages are required. In principle, however, high purity and yields of both components can be achieved since no eutectic is present.

If the impurity or minor component is completely or partially soluble in the solid phase of the component being purified, it is convenient to define a distribution coefficient k, defined by Eq. (20-1):

$$k = C_s/C_\ell \tag{20-1}$$

TABLE 20-1 Comparison of Processes Involving Crystallization from the Melt

Processes	Approximate upper melting point, °C	Materials tested	Minimum purity level obtained, ppm, weight	Product form
Progressive freezing	1500	All types	1	Ingot
Zone melting				
Batch	3500	All types	0.01	Ingot
Continuous	500	SiI_4	100	Melt
Melt crystallization				
Continuous	300	Organic	10	Melt
Cyclic	300	Organic	10	Melt

Abbreviated from Zief and Wilcox, *Fractional Solidification*, Marcel Dekker, New York, 1967, p. 7.

TABLE 20-2 Comparison of Features of Melt Crystallization and Distillation

Distillation	Melt crystallization
Phase equilibria	
Both liquid and vapor phases are totally miscible.	Liquid phases are totally miscible; solid phases are not.
Conventional vapor/liquid equilibrium.	Eutectic system.
Neither phase is pure.	Solid phase is pure, except at eutectic point.
Separation factors are moderate and decrease as purity increases.	Partition coefficients are very high (theoretically, they can be infinite).
Ultrahigh purity is difficult to achieve.	Ultrahigh purity is easy to achieve.
No theoretical limit on recovery.	Recovery is limited by eutectic composition.
Mass-transfer kinetics	
High mass-transfer rates in both vapor and liquid phases.	Only moderate mass-transfer rate in liquid phase, zero in solid.
Close approach to equilibrium.	Slow approach to equilibrium; achieved in brief contact time. Included impurities cannot diffuse out of solid.
Adiabatic contact assures phase equilibrium.	Solid phase must be remelted and refrozen to allow phase equilibrium.
Phase separability	
Phase densities differ by a factor of 100–10,000:1.	Phase densities differ by only about 10%.
Viscosity in both phases is low.	Liquid phase viscosity moderate, solid phase rigid.
Phase separation is rapid and complete.	Phase separation is slow; surface-tension effects prevent completion.
Countercurrent contacting is quick and efficient.	Countercurrent contacting is slow and imperfect.

Wynn, *Chem. Eng. Prog.*, **88**, 55 (1992). Reprinted with permission of the American Institute of Chemical Engineers. Copyright © 1992 AIChE. All rights reserved

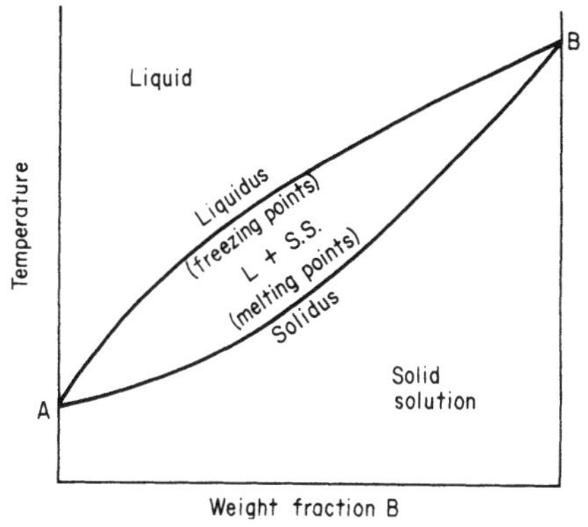

FIG. 20-1 Phase diagram for components exhibiting complete solid solution. (*Zief and Wilcox,* Fractional Solidification, *vol. 1, Marcel Dekker, New York, 1967, p. 31.*)

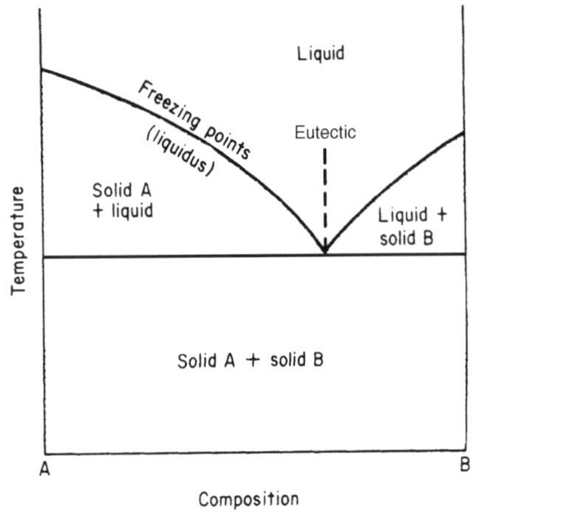

FIG. 20-2 Simple eutectic-phase diagram at constant pressure. (*Zief and Wilcox,* Fractional Solidification, *vol. 1, Marcel Dekker, New York, 1967, p. 24.*)

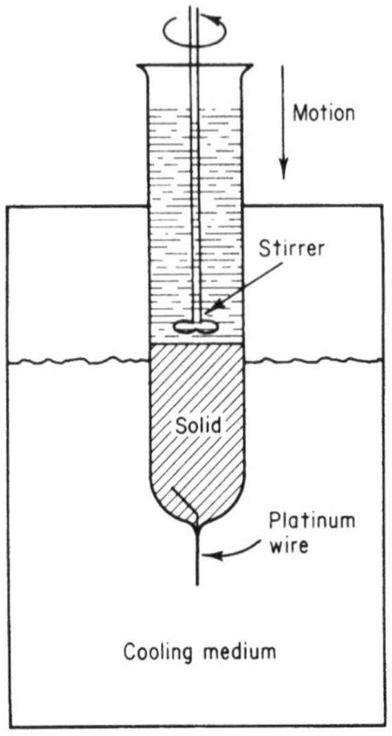

FIG. 20-3 Progressive freezing apparatus.

C_s is the concentration of impurity or minor component in the solid phase, and C_ℓ is the impurity concentration in the liquid phase. The distribution coefficient generally varies with composition. The value of k is greater than 1 when the solute raises the melting point and less than 1 when the melting point is depressed. In the regions near pure A or B the liquidus and solidus lines become linear; i.e., the distribution coefficient becomes constant. This is the basis for the common assumption of constant k in many mathematical treatments of fractional solidification in which ultrapure materials are obtained.

In the case of a simple eutectic system shown in Fig. 20-2, a pure solid phase is obtained by cooling if the composition of the feed mixture is not at the eutectic composition. If liquid composition is eutectic, then separate crystals of both species will form. In practice it is difficult to attain perfect separation of one component by crystallization of a eutectic mixture. The solid phase will always contain trace amounts of impurity because of incomplete solid-liquid separation, slight solubility of the impurity in the solid phase, or volumetric inclusions. It is difficult to generalize on which of these mechanisms is the major cause of contamination because of analytical difficulties in the ultrahigh-purity range.

The distribution-coefficient concept is commonly applied to fractional solidification of eutectic systems in the ultrapure portion of the phase diagram. If the quantity of impurity entrapped in the solid phase for whatever reason is proportional to that contained in the melt, then assumption of a constant k is valid. It should be noted that the theoretical yield of a component exhibiting binary eutectic behavior is fixed by the feed composition and position of the eutectic. Also, in contrast to the case of a solid solution, only one component can be obtained in a pure form.

There are many types of phase diagrams in addition to the two cases presented here; these are summarized in detail by Zief and Wilcox (op. cit., p. 21). Solid-liquid phase equilibria must be determined experimentally for most binary and multicomponent systems. Predictive methods are based mostly on ideal phase behavior and have limited accuracy near eutectics. A predictive technique based on extracting liquid-phase activity coefficients from *vapor-liquid* equilibria that is useful for estimating nonideal binary or multicomponent *solid-liquid* phase behavior has been reported by Muir (Pap. 71f, 73d ann. meet., AIChE, Chicago, 1980).

PROGRESSIVE FREEZING

Progressive freezing, sometimes called *normal* freezing, is the slow, directional solidification of a melt. Basically, this involves slow solidification at the bottom or sides of a vessel or tube by indirect cooling. The impurity is rejected into the liquid phase by the advancing solid interface. This technique can be employed to concentrate an impurity or, by repeated solidifications and liquid rejections, to produce a very pure ingot. Figure 20-3 illustrates a progressive freezing apparatus. The solidification rate and interface position are controlled by the rate of movement of the tube and the temperature of the cooling medium. There are many variations of the apparatus; e.g., the residual-liquid portion can be agitated and the directional freezing can be carried out vertically as shown in Fig. 20-3 or horizontally (see Richman et al., in Zief and Wilcox, op. cit., p. 259). In general, there is a solute redistribution when a mixture of two or more components is directionally frozen.

Component Separation by Progressive Freezing When the distribution coefficient is less than 1, the first solid which crystallizes contains less solute than the liquid from which it was formed. As the fraction which is frozen increases, the concentration of the impurity in the remaining liquid is increased and hence the concentration of impurity in the solid phase increases (for $k < 1$). The concentration gradient is reversed for $k > 1$. Consequently, in the absence of diffusion in the solid phase a concentration gradient is established in the frozen ingot.

One extreme of progressive freezing is equilibrium freezing. In this case the freezing rate must be slow enough to permit diffusion in the solid phase to eliminate the concentration gradient. When this occurs, there is no separation if the entire tube is solidified. Separation can be achieved, however, by terminating the freezing before all the liquid has been solidified. Equilibrium freezing is rarely achieved in practice because the diffusion rates in the solid phase are usually negligible (Pfann, op. cit., p. 10).

If the bulk-liquid phase is well mixed and no diffusion occurs in the solid phase, a simple expression relating the solid-phase composition to the fraction frozen can be obtained for the case in which the distribution coefficient is independent of composition and fraction frozen [Pfann, *Trans. Am. Inst. Mech. Eng.,* **194,** 747 (1952)].

$$C_s = kC_0(1 - X)^{k-1} \tag{20-2}$$

C_0 is the solution concentration of the initial charge, and X is the fraction frozen. Figure 20-4 illustrates the solute redistribution predicted by Eq. (20-2) for various values of the distribution coefficient.

There have been many modifications of this idealized model to account for variables such as the freezing rate and the degree of mixing in the liquid phase. For example, Burton et al. [*J. Chem. Phys.,* **21,** 1987 (1953)] reasoned that the solid rejects solute faster than it can diffuse into the bulk liquid. They proposed that the effect of the

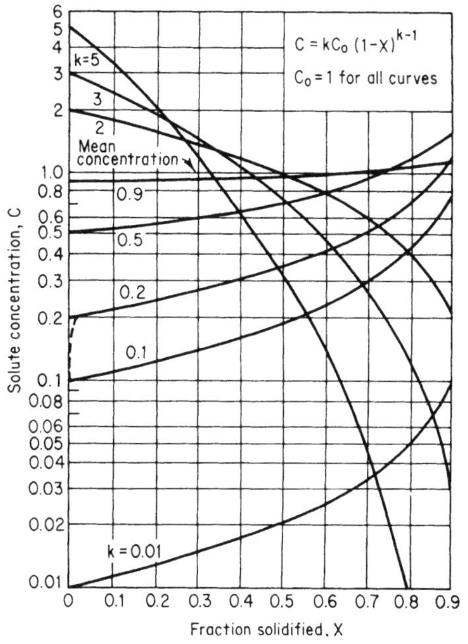

FIG. 20-4 Curves for progressive freezing, showing solute concentration C in the solid versus fraction-solidified X. (*Pfann,* Zone Melting, *2d ed., Wiley, New York, 1966, p. 12.*)

freezing rate and stirring could be explained by the diffusion of solute through a stagnant film next to the solid interface. Their theory resulted in an expression for an effective distribution coefficient k_{eff} which could be used in Eq. (20-2) instead of k.

$$k_{\text{eff}} = \frac{1}{1 + (1/k - 1)e^{-f\delta/D}} \tag{20-3}$$

where f = crystal growth rate, cm/s; δ = stagnant film thickness, cm; and D = diffusivity, cm²/s. No further attempt is made here to summarize the various refinements of Eq. (20-2). Zief and Wilcox (op. cit., p. 69) have summarized several of these models.

Pertinent Variables in Progressive Freezing The dominant variables which affect solute redistribution are the degree of mixing in the liquid phase and the rate of solidification. It is important to attain sufficient mixing to facilitate diffusion of the solute away from the solid-liquid interface to the bulk liquid. The film thickness δ decreases as the level of agitation increases. Cases have been reported in which essentially no separation occurred when the liquid was not stirred. The freezing rate which is controlled largely by the lowering rate of the tube (see Fig. 20-3) has a pronounced effect on the separation achieved. The separation is diminished as the freezing rate is increased. Also fluctuations in the freezing rate caused by mechanical vibrations and variations in the temperature of the cooling medium can decrease the separation.

Applications Progressive freezing has been applied to both solid solution and eutectic systems. As Fig. 20-4 illustrates, large separation factors can be attained when the distribution coefficient is favorable. Relatively pure materials can be obtained by removing the desired portion of the ingot. Also in some cases progressive freezing provides a convenient method of concentrating the impurities; e.g., in the case of $k < 1$ the last portion of the liquid that is frozen is enriched in the distributing solute.

Progressive freezing has been applied on the commercial scale. For example, aluminum has been purified by continuous progressive freezing [Dewey, *J. Metals*, **17,** 940 (1965)]. The Proabd refiner described by Molinari (Zief and Wilcox, op. cit., p. 393) is also a commercial example of progressive freezing. In this apparatus the mixture is directionally solidified on cooling tubes. Purification is achieved because the impure fraction melts first; this process is called sweating. This technique has been applied to the purification of naphthalene and p-dichlorobenzene and commercial equipment is available from BEFS PROKEM, Houston, Tx.

ZONE MELTING

Zone melting also relies on the distribution of solute between the liquid and solid phases to effect a separation. In this case, however, one or more liquid zones are passed through the ingot. This extremely versatile technique, which was invented by W. G. Pfann, has been used to purify hundreds of materials. Zone melting in its simplest form is illustrated in Fig. 20-5. A molten zone can be passed through an ingot from one end to the other by either a moving heater or by slowly drawing the material to be purified through a stationary heating zone.

Progressive freezing can be viewed as a special case of zone melting. If the zone length were equal to the ingot length and if only one pass were used, the operation would become progressive freezing. In general, however, when the zone length is only a fraction of the ingot length, zone melting possesses the advantage that a portion of the ingot does not have to be discarded after each solidification. The last portion of the ingot which is frozen in progressive freezing must be discarded before a second freezing.

Component Separation by Zone Melting The degree of solute redistribution achieved by zone melting is determined by the zone length l, ingot length L, number of passes n, the degree of mixing in the liquid zone, and the distribution coefficient of the materials being purified. The distribution of solute after one pass can be obtained by material-balance considerations. This is a two-domain problem; i.e., in the major portion of the ingot of length $L - l$ zone melting occurs in the conventional sense. The trailing end of the ingot of length l undergoes progressive freezing. For the case of constant-distribution coefficient, perfect mixing in the liquid phase, and negligible diffusion in the solid phase, the solute distribution for a single pass is given by Eq. (20-4) [Pfann, *Trans. Am. Inst. Mech. Eng.*, **194,** 747 (1952)].

$$C_s = C_0[1 - (1 - k)e^{-kx/\ell}] \tag{20-4}$$

The position of the zone x is measured from the leading edge of the ingot. The distribution for multiple passes can also be calculated from a material balance, but in this case the leading edge of the zone encounters solid corresponding to the composition at the point in question for the previous pass. The multiple-pass distribution has been numerically calculated (Pfann, *Zone Melting,* 2d ed., Wiley, New York, 1966, p. 285) for many combinations of k, L/l, and n. Typical solute-composition profiles are shown in Fig. 20-6 for various numbers of passes.

The ultimate distribution after an infinite number of passes is also shown in Fig. 20-6 and can be calculated for $x < (L - l)$ from the following equation (Pfann, op. cit., p. 42):

$$C_s = Ae^{BX} \tag{20-5}$$

where A and B can be determined from the following relations:

$$k = B\ell/(e^{B\ell} - 1) \tag{20-6}$$

$$A = C_0BL/(e^{BL} - 1) \tag{20-7}$$

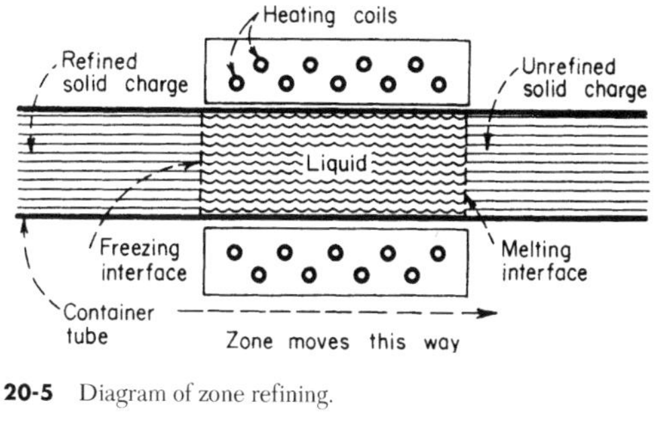

FIG. 20-5 Diagram of zone refining.

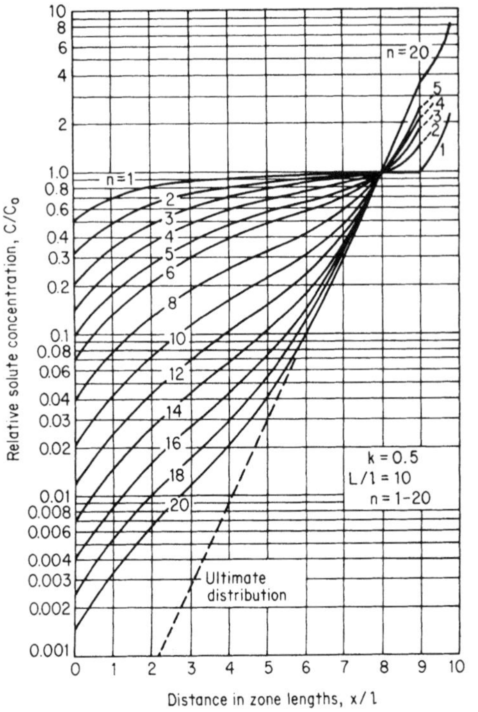

FIG. 20-6 Relative solute concentration C/C_0 (logarithmic scale) versus distance in zone lengths x/ℓ from beginning of charge, for various numbers of passes n. L denotes charge length. (*Pfann,* Zone Melting, *2d ed., Wiley, New York, 1966, p. 290.*)

The ultimate distribution represents the maximum separation that can be attained without cropping the ingot. Equation (20-5) is approximate because it does not include the effect of progressive freezing in the last zone length.

As in progressive freezing, many refinements of these models have been developed. Corrections for partial liquid mixing and a variable distribution coefficient have been summarized in detail (Zief and Wilcox, op. cit., p. 47).

Pertinent Variables in Zone Melting The dominant variables in zone melting are the number of passes, ingot-length–zone-length ratio, freezing rate, and degree of mixing in the liquid phase. Figure 20-6 illustrates the increased solute redistribution that occurs as the number of passes increases. Ingot-length–zone-length ratios of 4 to 10 are commonly used (Zief and Wilcox, op. cit., p. 624). An exception is encountered when one pass is used. In this case the zone length should be equal to the ingot length; i.e., progressive freezing provides the maximum separation when only one pass is used.

The freezing rate and degree of mixing have effects in solute redistribution similar to those discussed for progressive freezing. Zone travel rates of 1 cm/h for organic systems, 2.5 cm/h for metals, and 20 cm/h for semiconductors are common. In addition to the zone-travel rate the heating conditions affect the freezing rate. A detailed summary of heating and cooling methods for zone melting has been outlined by Zief and Wilcox (op. cit., p. 192). Direct mixing of the liquid region is more difficult for zone melting than progressive freezing. Mechanical stirring complicates the apparatus and increases the probability of contamination from an outside source. Some mixing occurs because of natural convection. Methods have been developed to stir the zone magnetically by utilizing the interaction of a current and magnetic field (Pfann, op. cit., p. 104) for cases in which the charge material is a reasonably good conductor.

Applications Zone melting has been used to purify hundreds of inorganic and organic materials. Many classes of inorganic compounds including semiconductors, intermetallic compounds, ionic salts, and oxides have been purified by zone melting. Organic materials of many types have been zone-melted. Zief and Wilcox (op. cit., p. 624) have compiled tables which give operating conditions and references for both inorganic and organic materials with melting points ranging from −115°C to over 3000°C.

Some materials are so reactive that they cannot be zone-melted to a high degree of purity in a container. Floating-zone techniques in which the molten zone is held in place by its own surface tension have been developed by Keck et al. [*Phys. Rev.*, **89,** 1297 (1953)].

Continuous-zone-melting apparatus has been described by Pfann (op. cit., p. 171). This technique offers the advantage of a close approach to the ultimate distribution, which is usually impractical for batch operation.

Performance data have been reported by Kennedy et al. (*The Purification of Inorganic and Organic Materials,* Marcel Dekker, New York, 1969, p. 261) for continuous-zone refining of benzoic acid.

MELT CRYSTALLIZATION FROM THE BULK

Conducting crystallization inside a vertical or horizontal column with a countercurrent flow of crystals and liquid can produce a higher product purity than conventional crystallization or distillation. The working concept is to form a crystal phase from the bulk liquid, either internally or externally, and then transport the solids through a countercurrent stream of enriched reflux liquid obtained from melted product. The problem in practicing this technology is the difficulty of controlling solid-phase movement. Unlike distillation, which exploits the specific-gravity differences between liquid and vapor phases, melt crystallization involves the contacting of liquid and solid phases that have nearly identical physical properties. Phase densities are frequently very close, and gravitational settling of the solid phase may be slow and ineffective. The challenge of designing equipment to accomplish crystallization in a column has resulted in a myriad of configurations to achieve reliable solid-phase movement, high product yield and purity, and efficient heat addition and removal.

Investigations Crystallization conducted inside a column is categorized as either **end-fed** or **center-fed** depending on whether the feed location is upstream or downstream of the crystal forming section. Figure 20-7 depicts the features of an end-fed commercial column described by McKay et al. [*Chem. Eng. Prog. Symp. Ser.,* no. 25, 55, 163 (1969)] for the separation of xylenes. Crystals of p-xylene are formed by indirect cooling of the melt in scraped-surface heat exchangers, and the resultant slurry is introduced into the column at the top. This type of column has no mechanical internals to transport solids and instead relies upon an imposed hydraulic gradient to force the solids through the column into the melting zone. Residue liquid is removed through a filter directly above the melter. A pulse piston in the product discharge improves washing efficiency and column reliability.

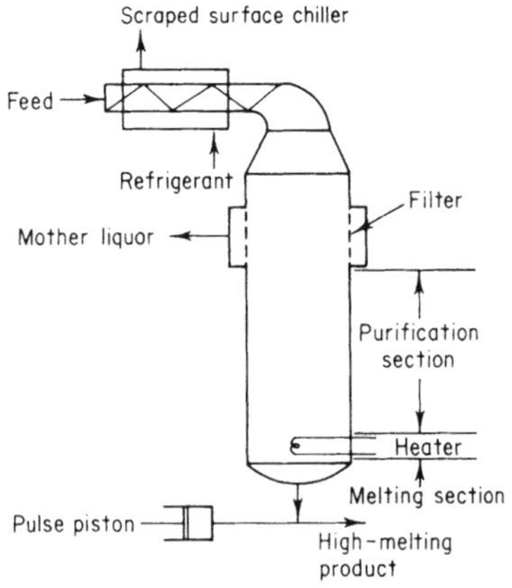

FIG. 20-7 End-fed column crystallizer. (*Phillips Petroleum Co.*)

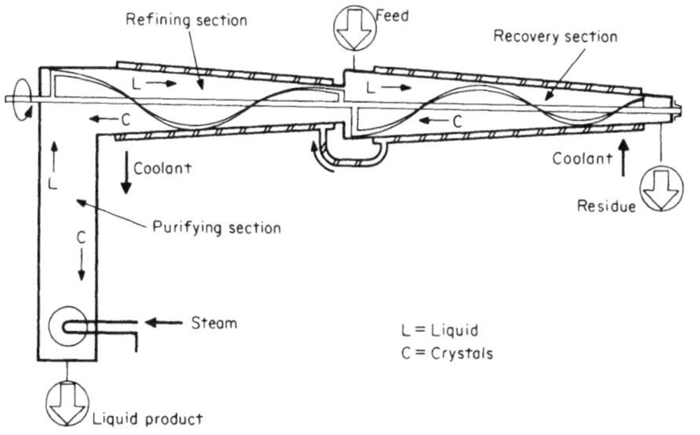

FIG. 20-8 Horizontal center-fed column crystallizer. (*The C. W. Nofsinger Co.*)

Figure 20-8 shows the features of a horizontal center-fed column [Brodie, *Aust. Mech. Chem. Eng. Trans.*, **37** (May 1979)] which has been commercialized for continuous purification of naphthalene and *p*-dichlorobenzene. Liquid feed enters the column *between* the hot purifying section and the cold freezing or recovery zone. Crystals are formed internally by indirect cooling of the melt through the walls of the refining and recovery zones. Residue liquid that has been depleted of product exits from the coldest section of the column. A spiral conveyor controls the transport of solids through the unit.

Another center-fed design that has only been used on a preparative scale is the vertical spiral conveyor column reported by Schildknecht [*Angew. Chem.*, **73,** 612 (1961)]. In this device, a version of which is shown on Fig. 20-9, the dispersed-crystal phase is formed in the freezing section and conveyed downward in a controlled manner by a rotating spiral with or without a vertical oscillation.

Differences have been observed in the performance of end- and center-fed column configurations. Consequently, discussions of center- and end-fed column crystallizers are presented separately. The design and operation of both columns are reviewed by Powers (Zief and Wilcox, op. cit., p. 343). A comparison of these devices is shown on Table 20-3.

Center-Fed Column Crystallizers Two types of center-fed column crystallizers are illustrated on Figs. 20-8 and 20-9. As in a simple distillation column, these devices are composed of three distinct sections: a freezing or recovery section, where solute is frozen from the impure liquor; the purification zone, where countercurrent contacting of solids and liquid occurs; and the crystal-melting and -refluxing section. Feed position separates the refining and recovery portions of the purification zone. The section between feed location and melter is referred to as the refining or enrichment section, whereas the section between feed addition and freezing is called the recovery section. The refining section may have provisions for sidewall cooling. The published literature on column crystallizers connotes stripping and refining in a reverse sense to distillation terminology, since refined product from a melt crystallizer exits at the hot section of the column rather than at the cold end as in a distillation column.

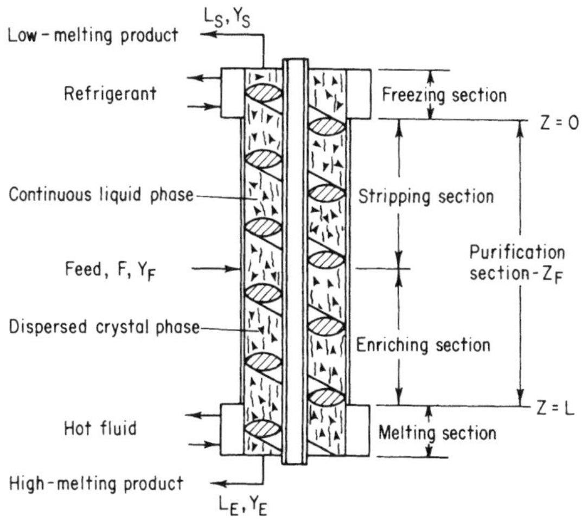

FIG. 20-9 Center-fed column crystallizer with a spiral-type conveyor.

TABLE 20-3 Comparison of Melt-Crystallizer Performance

Center-fed column	End-fed column
Solid phase is formed internally; thus, only liquid streams enter and exit the column.	Solid phase is formed in external equipment and fed as slurry into the purifier.
Internal reflux can be controlled without affecting product yield.	The maximum internal liquid reflux is fixed by the thermodynamic state of the feed relative to the product stream. Excessive reflux will diminish product yield.
Operation can be continuous or batchwise at total reflux.	Total reflux operation is not feasible.
Center-fed columns can be adapted for both eutectic and solid-solution systems.	End-fed columns are inefficient for separation of solid-solution systems.
Either low- or high-porosity-solids-phase concentrations can be formed in the purification and melting zones.	End-fed units are characterized by low-porosity-solids packing in the purification and melting zones.
Scale-up depends on the mechanical complexity of the crystal-transport system and techniques for removing heat. Vertical oscillating spiral columns are likely limited to about 0.2 m in diameter, whereas horizontal columns of several meters are possible.	Scale-up is limited by design of melter and/or crystal-washing section. Vertical or horizontal columns of several meters in diameter are possible.

Rate processes that describe the purification mechanisms in a column crystallizer are highly complex since phase transition and heat- and mass-transfer processes occur simultaneously. Nucleation and growth of a crystalline solid phase along with crystal washing and crystal melting are occurring in various zones of the apparatus. Column hydrodynamics are also difficult to describe. Liquid- and solid-phase mixing patterns are influenced by factors such as solids-transport mechanism, column orientation, and, particularly for dilute slurries, the settling characteristics of the solids.

Most investigators have focused their attention on a differential segment of the zone between the feed injection and the crystal melter. Analysis of crystal formation and growth in the recovery section has received scant attention. Table 20-4 summarizes the scope of the literature treatment for center-fed columns for both solid-solution and eutectic forming systems.

The dominant mechanism of purification for column crystallization of solid-solution systems is recrystallization. The rate of mass transfer resulting from recrystallization is related to the concentrations of the solid phase and free liquid which are in intimate contact. A model based on height-of-transfer-unit (HTU) concepts representing the composition profile in the purification section for the high-melting component of a binary solid-solution system has been reported by Powers et al. (in Zief and Wilcox, op. cit., p. 363) for total-reflux operation. Typical data for the purification of a solid-solution system, azobenzene-stilbene, are shown in Fig. 20-10. The column crystallizer was operated at total reflux. The solid line through the data was computed by Powers et al. (op. cit., p. 364) by using an experimental HTU value of 3.3 cm.

TABLE 20-4 Column-Crystallizer Investigations

	Treatments	
	Theoretical	Experimental
Solid solutions		
Total reflux—steady state	1, 2, 4, 6	1, 4, 6
Total reflux—dynamic	2	
Continuous—steady state	1, 4	4, 8, 9
Continuous—dynamic		
Eutectic systems		
Total reflux—steady state	1, 3, 4, 7	1, 3, 6
Total reflux—dynamic		
Continuous—steady state	1, 5, 10, 11, 12	5, 8, 9, 10, 11, 13
Continuous—dynamic		

1. Powers, *Symposium on Zone Melting and Column Crystallization,* Karlsruhe, 1963.
2. Anikin, *Dokl. Akad. Nauk SSSR,* **151,** 1139 (1969).
3. Albertins et al., *Am. Inst. Chem. Eng. J.,* **15,** 554 (1969).
4. Gates et al., *Am. Inst. Chem. Eng. J.,* **16,** 648 (1970).
5. Henry et al., *Am. Inst. Chem. Eng. J.,* **16,** 1055 (1970).
6. Schildknecht et al., *Angew. Chem.,* **73,** 612 (1961).
7. Arkenbout et al., *Sep. Sci.,* **3,** 501 (1968).
8. Betts et al., *Appl. Chem.,* **17,** 180 (1968).
9. McKay et al., *Chem. Eng. Prog. Symp. Ser.,* no. 25, 55, 163 (1959).
10. Bolsaitis, *Chem. Eng. Sci.,* **24,** 1813 (1969).
11. Moyers et al., *Am. Inst. Chem. Eng. J.,* **20,** 1119 (1974).
12. Griffin, M.S. thesis in chemical engineering, University of Delaware, 1975.
13. Brodie, *Aust. Mech. Chem. Eng. Trans.,* 37 (1971).

Most of the analytical treatments of center-fed columns describe the purification mechanism in an adiabatic oscillating spiral column (Fig. 20-9). However, the analyses by Moyers (op. cit.) and Griffin (op. cit.) are for a nonadiabatic dense-bed column. Differential treatment of the horizontal-purifier (Fig. 20-8) performance has not been reported; however, overall material and enthalpy balances have been described by Brodie (op. cit.) and apply equally well to other designs.

A dense-bed center-fed column (Fig. 20-11) having provision for internal crystal formation and variable reflux was tested by Moyers et al. (op. cit.). In the theoretical development (ibid.) a nonadiabatic, plug-flow axial-dispersion model was employed to describe the performance of the entire column. Terms describing interphase transport of impurity between adhering and free liquid are not considered.

A comparison of the axial-dispersion coefficients obtained in oscillating-spiral and dense-bed crystallizers is given in Table 20-5. The dense-bed column approaches axial-dispersion coefficients similar to those of densely packed ice-washing columns.

The concept of minimum reflux as related to column-crystallizer operation is presented by Brodie (op. cit.) and is applicable to all types of column crystallizers, including end-fed units. In order to stabilize column operation the sensible heat of subcooled solids entering the melting zone should be balanced or exceeded by the heat of fusion of the refluxed melt. The relationship in Eq. (20-8) describes the minimum reflux requirement for proper column operation.

$$R = (T_P - T_F)\, C_P/\lambda \tag{20-8}$$

R = reflux ratio, g reflux/g product; T_P = product temperature, °C; T_F = saturated-feed temperature, °C; C_P = specific heat of solid crystals, cal/(g·°C); and λ = heat of fusion, cal/g.

All refluxed melt will refreeze if reflux supplied equals that computed by Eq. (20-8). When reflux supplied is greater than the minimum, jacket cooling in the refining zone or additional cooling in the

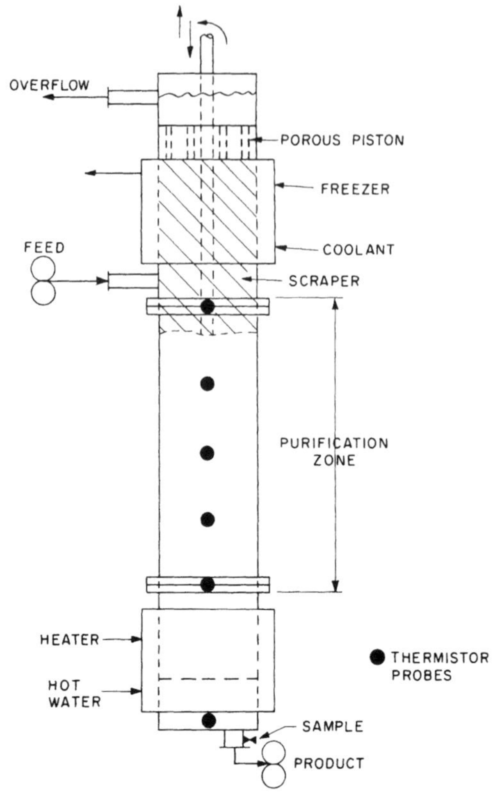

FIG. 20-11 Dense-bed center-fed column crystallizer. [*Moyers et al.,* Am. Inst. Chem. Eng. J., ***20,*** *1121 (1974).*]

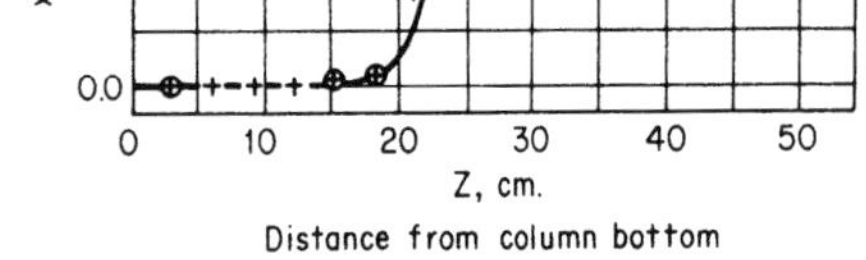

FIG. 20-10 Steady-state separation of azobenzene and stilbene in a center-fed column crystallizer with total-reflux operation. To convert centimeters to inches, multiply by 0.3937. (*Zief and Wilcox,* Fractional Solidification, *vol. 1, Marcel Dekker, New York, 1967, p. 356.*)

TABLE 20-5 Comparison of Axial-Dispersion Coefficients for Several Liquid-Solid Contactors

Column type	Dispersion coefficient, cm²/s	Reference
Center-fed crystallizer (oscillating spiral)	1.6–3.5	1
Center-fed crystallizer (oscillating spiral)	1.3–1.7	2
Countercurrent ice-washing column	0.025–0.17	3
Center-fed crystallizer	0.12–0.30	4

References:
1. Albertins et al., *Am. Inst. Chem. Eng. J.,* **15,** 554 (1969).
2. Gates et al., *Am. Inst. Chem. Eng. J.,* **16,** 648 (1970).
3. Ritter, Ph.D. thesis, Massachusetts Institute of Technology, 1969.
4. Moyers et al., *Am. Inst. Chem. Eng. J.,* **20,** 1119 (1974).

recovery zone is required to maintain product recovery. Since high-purity melts are fed near their pure-component freezing temperatures, little refreezing takes place unless jacket cooling is added.

To utilize a column-crystallizer design or rating model, a large number of parameters must be identified. Many of these are empirical in nature and must be determined experimentally in equipment identical to the specific device being evaluated. Hence macroscopic evaluation of systems by large-scale piloting is the rule rather than the exception. Included in this rather long list of critical parameters are factors such as impurity level trapped in the solid phase, product quality as a function of reflux ratio, degree of liquid and solids axial mixing in the equipment as a function of solids-conveyor design, size and shape of crystals produced, and ease of solids handling in the column. Heat is normally removed through metal surfaces; thus, the stability of the solution to subcooling can also be a major factor in design.

End-Fed Column Crystallizer End-fed columns were developed and successfully commercialized by the Phillips Petroleum Company in the 1950s. The sections of a typical end-fed column, often referred to as a Phillips column, are shown on Fig. 20-7. Impure liquor is removed through filters located between the product-freezing zone and the melter rather than at the end of the freezing zone, as occurs in center-fed units. The purification mechanism for end-fed units is basically the same as for center-fed devices. However, there are reflux restrictions in an end-fed column, and a high degree of solids compaction exists near the melter of an end-fed device. It has been observed that the free-liquid composition and the fraction of solids are relatively constant throughout most of the purification section but exhibit a sharp discontinuity near the melting section [McKay et al., *Ind. Eng. Chem.*, **52,** 197 (1969)]. Investigators of end-fed column behavior are listed in Table 20-6. Note that end-fed columns are adaptable only for eutectic-system purification and cannot be operated at total reflux.

Performance information for the purification of *p*-xylene indicates that nearly 100 percent of the crystals in the feed stream are removed as product. This suggests that the liquid which is refluxed from the melting section is effectively refrozen by the countercurrent stream of subcooled crystals. A high-melting product of 99.0 to 99.8 weight percent *p*-xylene has been obtained from a 65 weight percent *p*-xylene feed. The major impurity was *m*-xylene. Figure 20-12 illustrates the column-cross-section-area–capacity relationship for various product purities.

Column crystallizers of the end-fed type can be used for purification of many eutectic-type systems and for aqueous as well as organic systems (McKay, loc. cit.). Column crystallizers have been used for xylene isomer separation, but recently other separation technologies including more efficient melt crystallization equipment have tended to supplant the Phillips style crystallizer.

Commercial Equipment and Applications In the last two decades the practice of melt crystallization techniques for purification of certain organic materials has made significant commercial progress. The concept of refining certain products by countercurrent staging of crystallization in a column has completed the transition from laboratory and pilot equipment to large-scale industrial configurations. Chemicals which have been purified by suspension crystallization-purifier column techniques are listed on Table 20-7. The practice of crystal formation and growth from the bulk liquid (as is practiced in suspension crystallization techniques described in Sec. 18 of this handbook) and subsequent crystal melting and refluxing in a purifier column has evolved into two slightly different concepts: (1) the horizontal continuous crystallization technique with vertical purifier invented by Brodie (op. cit.) and (2) the continuous multistage or stepwise system with vertical purifier developed by Tsukishima Kikai Co., Ltd. (TSK). A recent description of these processes has been published by Meyer [*Chem. Proc.*, **53,** 50 (1990)].

The horizontal continuous Brodie melt crystallizer is basically an indirectly cooled crystallizer with an internal ribbon conveyor to transport crystals countercurrent to the liquid and a vertical purifier for final refining. Figure 20-8 describes the operation of a single tube unit and Fig. 20-13 depicts a multitube unit. The multitube design has been successfully commercialized for a number of organic chemicals. The Brodie purifier configuration requires careful control of process and equipment temperature differences to eliminate internal encrustations and is limited by the inherent equipment geometry to capacities of less than 15,000 tons per year per module.

In the multistage process described on Fig. 20-14 feed enters one of several crystallizers installed in series. Crystals formed in each crystallizer are transferred to a hotter stage and the liquid collected in the clarified zone of the crystallizer is transferred to a colder stage and eventually discharged as residue. At the hot end, crystals are transferred to a vertical purifier where countercurrent washing is performed by pure, hot-product reflux. TSK refers to this multistage process as the countercurrent cooling crystallization (CCCC) process. In

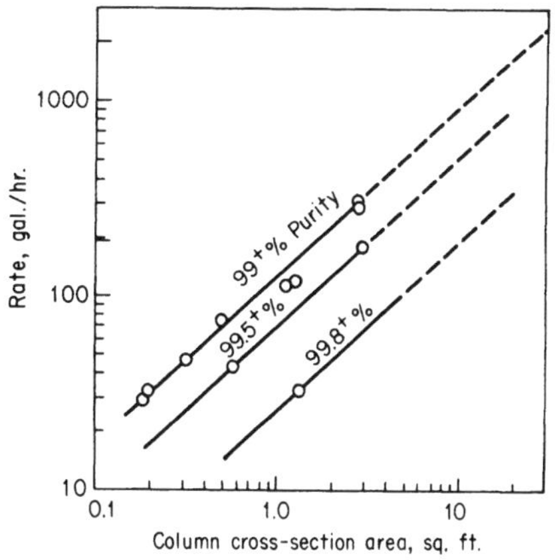

FIG. 20-12 Pulsed-column capacity versus column size for 65 percent *p*-xylene feed. To convert gallons per hour to cubic meters per hour, multiply by 0.9396; to convert square feet to square meters, multiply by 0.0929. (*McKay et al., prepr., 59th nat. meet. AIChE, East Columbus, Ohio.*)

TABLE 20-6 End-Fed-Crystallizer Investigations

	Treatments	
Eutectic systems	Theoretical	Experimental
Continuous—steady state	1, 2, 4	1, 4
Batch	3	3

1. McKay et al., *Ind. Eng. Chem. Process Des. Dev.*, **6,** 16 (1967).
2. Player, *Ind. Eng. Chem. Process Des. Dev.*, **8,** 210 (1969).
3. Yagi et al., *Kagaku Kogaku*, **72,** 415 (1963).
4. Shen and Meyer, Prepr. 19F, AIChE Symp., Chicago, 1970.

TABLE 20-7 Chemicals Purified by TSK CCCC Process (The C. W. Nofsinger Co.)

Acetic acid	Isophthaloyl chloride
Acrylic acid	Isopregol
Adipic acid	Lutidine
Benzene	Maleic anhydride
Biphenyl	Naphthalene
Bisphenol-A	*p*-Nitrochloro benzene
Caprolactam	*p*-Nitrotoluene
Chloroacetic acid	Phenol
p-Chloro toluene	*b*-Picoline
p-Cresol	*g*-Picoline
Combat (proprietary)	Pyridine
Dibutyl hydroxy toluene (BHT)	Stilbene
p-Dichloro benzene	Terephthaloyl chloride
2,5 Dichlorophenol	Tertiary butyl phenol
Dicumyl peroxide	Toluene diisocyanate
Diene	Trioxane
Heliotropin	*p*-Xylene
Hexachloro cyclo butene	3,4 Xylidine
Hexamethylene diamine	

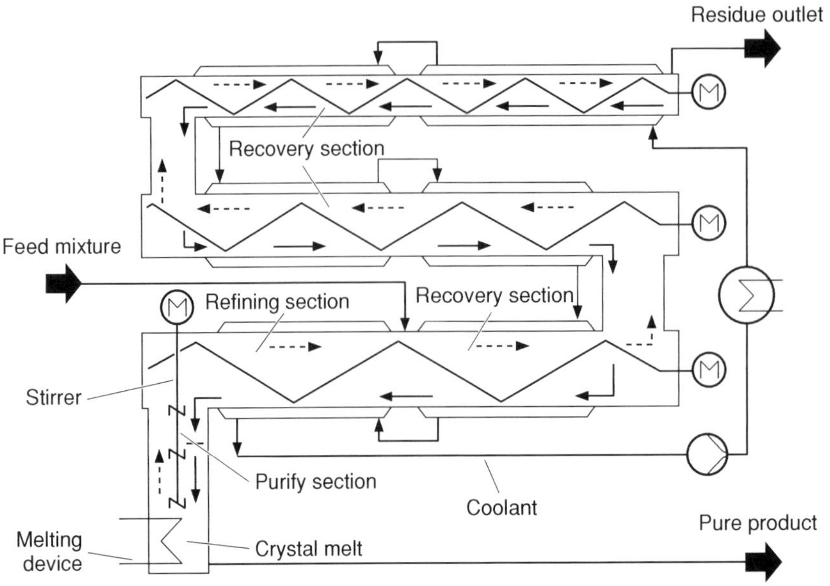

FIG. 20-13 Horizontal continuous Brodie melt crystallizer—multitube unit. (*The C. W. Nofisinger Co.*)

principle any suitable type of crystallizer can be used in the stages as long as the crystals formed can be separated from the crystallizer liquid and settled and melted in the purifier.

Commercial applications for both the Brodie and CCCC process are indicated on Table 20-8. Both the Brodie Purifier and the CCCC processes are available from The C. W. Nofsinger Company, PO Box 419173, Kansas City, MO 64141-0173.

FALLING-FILM CRYSTALLIZATION

Falling-film crystallization utilizes progressive freezing principles to purify melts and solutions. The technique established to practice the process is inherently cyclic. Figure 20-15 depicts the basic working concept. First a crystalline layer is formed by subcooling a liquid film on a vertical surface inside a tube. This coating is then

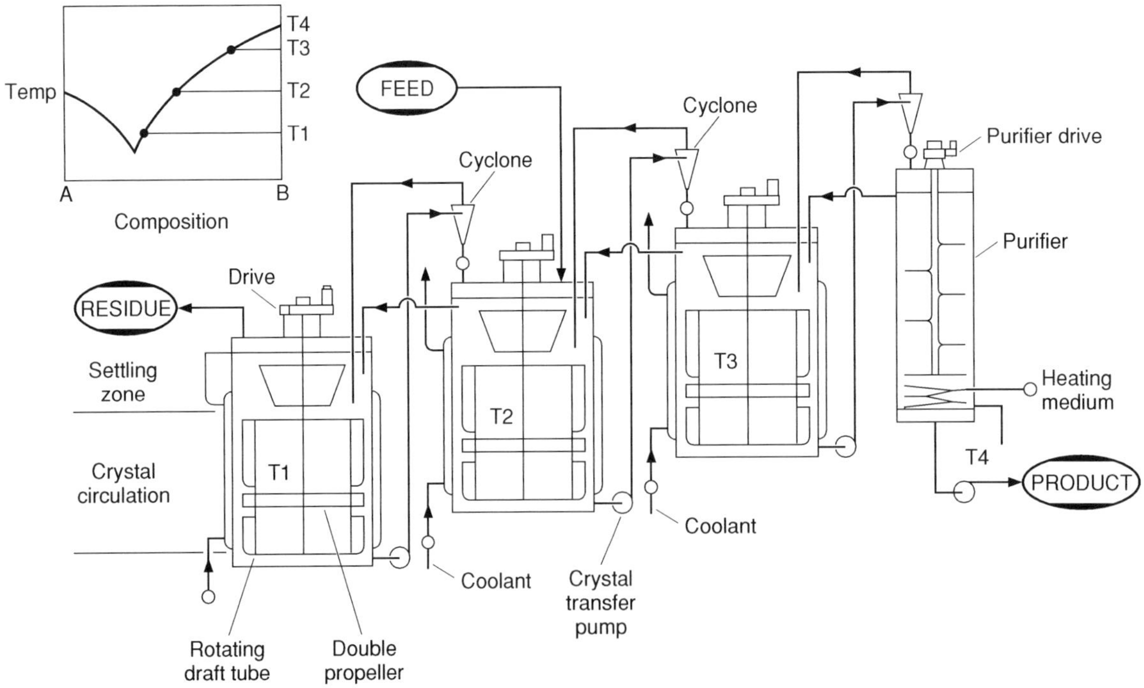

FIG. 20-14 Crystallizer—multistage process. (*The C. W. Nofisinger Co.*)

TABLE 20-8 Commercial TSK Crystallization Operating Plants

	Capacity MM lb/yr	Company	Date & location
	Countercurrent Cooling Crystallization (CCCC) Process		
Nofsinger license			
Nofsinger design & construct			
p-Dichlorobenzene	Confidential	Monsanto Co.	1989—Sauget, IL
TSK license			
TSK design & construct			
Confidential[1]	Confidential	Confidential	1988—Japan
p-Xylene	137	MGC	1986—Mizushima, Japan
Confidential[2]	Confidential	Confidential	1985—Japan
p-Xylene	132	MGC	1983—Mizushima, Japan
	expanded to 160		1984—Japan
p-Xylene	26.5	MGC	1981—Mizushima, Japan
	Brodie		
TSK license			
TSK design & construct			
Naphthalene	8	SHSM	1985—China
p-DCB	13	Hodogaya	1981—Japan
p-DCB	5.5	Sumitomo	1978—Japan
UCAL license & design			
TSK hardware			
Naphthalene	16	Nippon Steel	1974—Japan
p-DCB	13	Hodogaya	1974—Japan
UCAL license & design			
Naphthalene	10	British Tar	1972—U.K.
p-DCB	3	UCAL	1969—Australia

ABBREVIATIONS:
Nofsinger The C. W. Nofsinger Company
TSK Tsukishima Kikai Co., Ltd.
SHSM Shanghai Hozan Steel Mill
UCAL Union Carbide Australia, Ltd.
MGC Mitsubishi Gas Chemical Co., co-developer with TSK of the application for *p*-Xylene

1. Commercial scale plant started up in the spring of 1988 purifying a bulk chemical. This is the first application of the CCCC process on this bulk chemical.
2. This small unit is operating in Japan with an 800-mm crystallizer and 300-mm purifier. Because of confidentiality, we cannot disclose the company, capacity, or product.

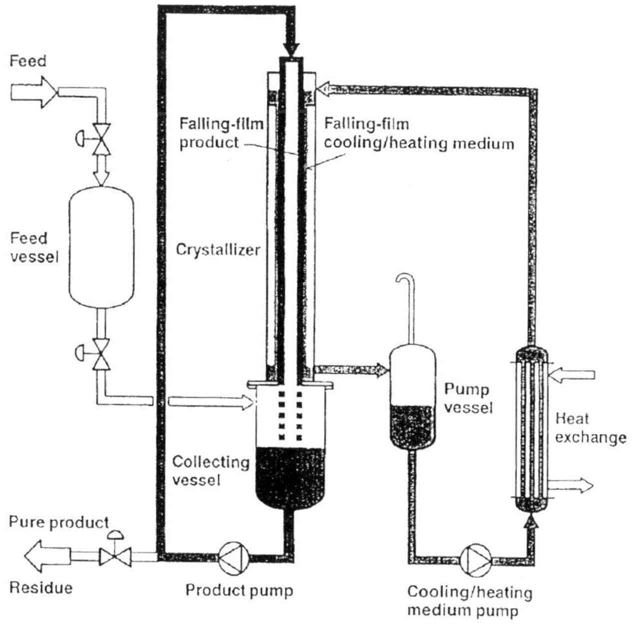

FIG. 20-15 Dynamic crystallization system. (*Sulzer Chemtech.*)

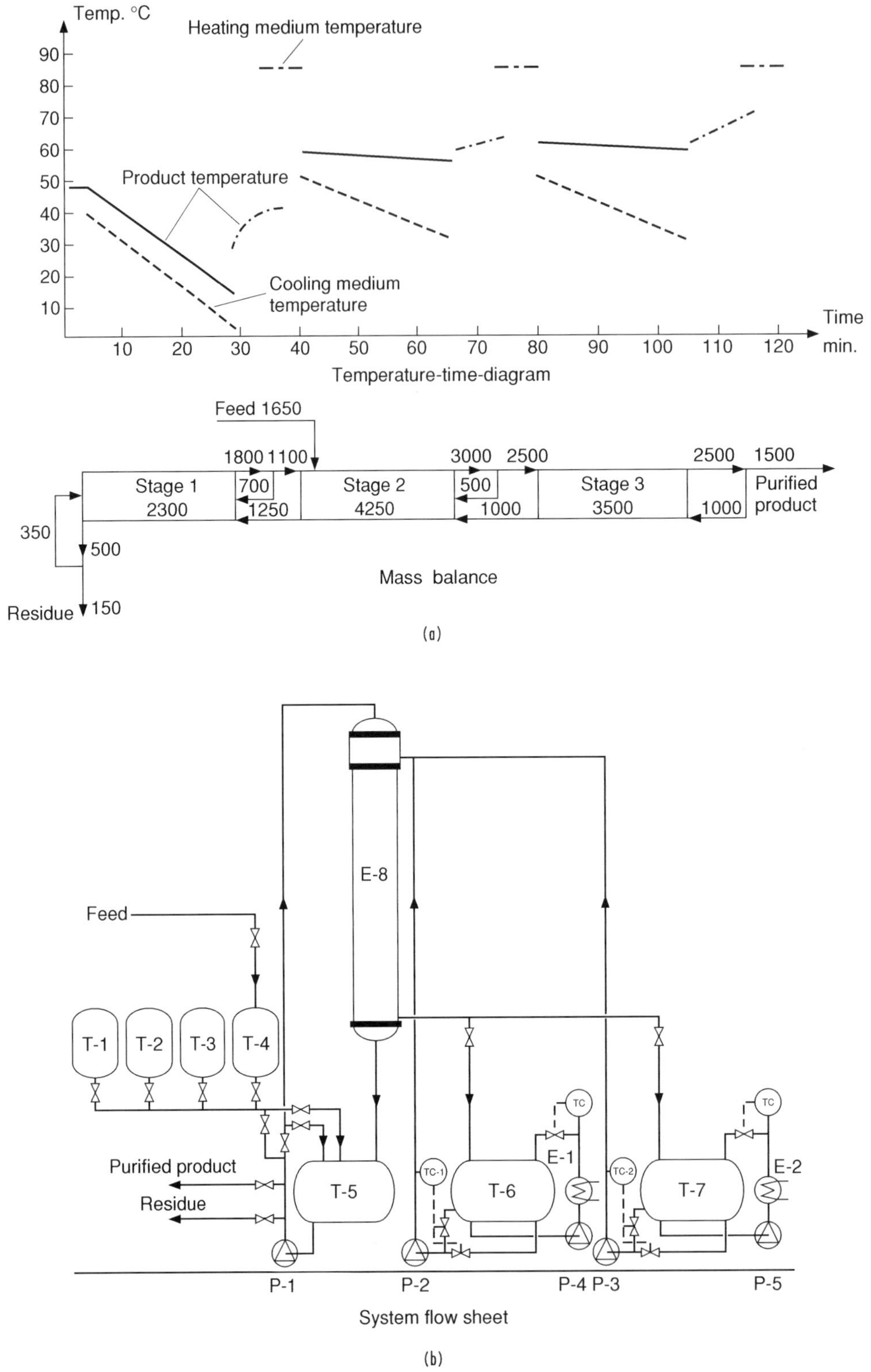

FIG. 20-16 Sulzer MWB-crystallization process. (*a*) Stepwise operation of the process. (*b*) System flow sheet. (*Sulzer Chemtech.*)

grown by extracting heat from a falling film of melt (or solution) through a heat transfer surface. Impure liquid is then drained from the crystal layer and the product is reclaimed by melting. Variants of this technique have been perfected and are used commercially for many types of organic materials. Both static and falling-film techniques have been described by Wynn [*Chem. Eng. Progr.*, (1992)]. Mathematical models for both static and dynamic operations have been presented by Gilbert [*AIChE J.*, **37,** 1205 (1991)].

Principles of Operation Figure 20-16 describes a typical three-stage falling-film crystallization process for purification of MCA (monochloro acetic acid). Crystallizer E-8 consists of a number of vertical tubular elements working in parallel enclosed in a shell. Normal tube length is 12 meters with a 50- to 75-millimeter tube inside diameter. Feed enters stage two of the sequential operation, is added to the kettle (T-5), and is then circulated to the top of the crystallizer and distributed as a falling film inside the tubes. Nucleation is induced at the inside walls and a crystal layer starts to grow. Temperature of the coolant is progressively lowered to compensate for reduced heat transfer and lower melt freezing point until the thickness inside the tube is between 5 and 20 millimeters depending on the product. Kettle liquid is evacuated to the first-stage holding tank (T-3) for eventual recrystallization at a lower temperature to maximize product yield and to strip product from the final liquid residue. Semirefined product frozen to the inside of the tube during operation of stage two is first heated above its melting point and slightly melted (sweated). This semipurified melted material (sweat) is removed from the crystallizer kettle, stored in a stage tank (T-4), and then added to the next batch of fresh feed. The remaining material inside the crystallizer is then melted, mixed with product sweat from stage three, recrystallized, and sweated to upgrade the purity even further (stage 3).

Commercial Equipment and Applications The falling-film crystallization process was invented by the MWB company in Switzerland. The process is now marketed by Sulzer Chemtech. Products successfully processed in the falling-film crystallizer are listed on Table 20-9. The falling-film crystallization process is available from the Chemtech Div. of Sulzer Canada Inc., 60 Worcester Rd., Rexdale, Ontario N9W 5X2 Canada.

TABLE 20-9 Fractional Crystallization Reference List

Product	Main characteristics	Capacity, tons/year	Purity	Type of plant	Country	Client
Acrylic acid	Very low aldehyde content, no undesired polymerization in the plant	Undisclosed Undisclosed	99.95% 99.9%	Falling film Falling film	Undisclosed Undisclosed	Undisclosed Undisclosed
Benzoic acid	Pharmaceutical grade, odor- and color free	4,500	99.97%	Falling film	Italy	Chimica del Friuli
Bisphenol A	Polycarbonate grade, no solvent required	150,000	Undisclosed	Falling film	USA	General Electric
Carbonic acid		1,200	Undisclosed	Falling film	Germany	Undisclosed
Fatty acid	Separation of tallow fatty acid into saturated and unsaturated fractions	20,000	Stearic acid: Iodine no. 2 Oleic acid: Cloud pt 5°C	Falling film	Japan	Undisclosed
Fine chemicals		<1,000 <1,000 <1,000 <1,000 <1,000 <1,000 <1,000	Undisclosed Undisclosed Undisclosed Undisclosed Undisclosed Undisclosed Undisclosed	Falling film Static Falling film Falling film Falling film Falling film Falling film	GUS Switzerland Switzerland Switzerland USA Germany Japan	Undisclosed Undisclosed Undisclosed Undisclosed Undisclosed Undisclosed Undisclosed
Hydrazine	Satellite grade	3	>99.9%	Falling film	Germany	ESA
Monochloro acetic acid (MCA)	Low DCA content	6,000	>99.2%	Falling film	USA	Undisclosed
Multipurpose	Separation or purification of two or more chemicals, alternatively	1,000 1,000	Various grades Undisclosed	Falling film Falling film	Belgium Belgium	UCB Reibelco
Naphthalene	Color free and color stable with low thionaphthene content	60,000 20,000 10,000 12,000	99.5% 99.5% 99.8% Various grades	Falling film Falling film/static Falling film/static Falling film	Germany P.R. China P.R. China The Netherlands	Rütgers-Werke Anshan Jining Cindu Chemicals
p-Dichlorobenzene	No solvent washing required	40,000 5,000 4,000 3,000 3,000	99.95% 99.98% 99.8% 99.95% >97%	Falling film Falling film Falling film/distillation Falling film Static	USA Japan Brazil P.R. China P.R. China	Standard Chlorine Toa Gosei Nitroclor Fuyang Shandong
p-Nitrochlorobenzene		18,000 10,000	99.3% 99.5%	Falling film/distillation Static	P.R. China India	Jilin Chemical Mardia
Toluene diisocyanate (TDI)	Separation of TDI 80 into TDI 100 & TDI 65	20,000	Undisclosed	Falling film	Undisclosed	Undisclosed
Trioxan		<1,000	99.97%	Falling film	Undisclosed	Undisclosed

SUPERCRITICAL FLUID SEPARATION PROCESSES

GENERAL REFERENCES: Yeo and Kiran, *J. Supercritical Fluids,* **34,** 287–308 (2005). York, Kompella, and Shekunov, *Supercritical Fluid Technology for Drug Product Development*, Marcel Dekker, New York, 2004. Shah, Hanrath, Johnston, and Korgel, *J. Physical Chemistry B,* **108,** 9574–9587 (2004). Eckert, Liotta, Bush, Brown, and Hallett, *J. Physical Chemistry B,* **108,** 18108–18118 (2004). DeSimone, *Science,* **297,** 799–803 (2002). Arai, Sako, and Takebayashi, *Supercritical Fluids: Molecular Interactions, Physical Properties, and New Applications*, Springer, New York, 2002. Kiran, Debenedetti, and Peters, *Supercritical Fluids: Fundamentals and Applications*, Kluwer Academic, Dordrecht, 2000. McHugh and Krukonis, *Supercritical Fluid Extraction Principles and Practice*, 2d ed., Butterworth, Stoneham, Mass., 1994. Brunner, *Gas Extraction: An Introduction to Fundamentals of Supercritical Fluids and the Application to Separation Processes*, Springer, New York, 1994. Gupta and Shim, *Solubility in Supercritical Carbon Dioxide*, CRC Press, Boca Raton, Fla., 2007. Gupta and Kompella, *Nanoparticle Technology for Drug Delivery*, Taylor & Francis, New York, 2006.

INTRODUCTION

Fluids above their critical temperatures and pressures, called supercritical fluids (SCFs), exhibit properties intermediate between those of gases and liquids. Consequently, each of these two boundary conditions sheds insight into the nature of these fluids. Unlike gases, SCFs possess a considerable solvent strength, and transport properties are more favorable. In regions where a SCF is highly compressible, its density and hence its solvent strength may be adjusted over a wide range with modest variations in temperature and pressure. This tunability may be used to control phase behavior, separation processes (e.g., SCF extraction), rates and selectivities of chemical reactions, and morphologies in materials processing. For SCF separation processes to be feasible, the advantages (Table 20-10) must compensate for the costs of high pressure; examples of commercial applications are listed in Table 20-11.

The two SCFs most often studied—CO_2 and water—are the two least expensive of all solvents. CO_2 is nontoxic and nonflammable and has a near-ambient critical temperature of 31.1°C. CO_2 is an environmentally friendly substitute for organic solvents including chlorocarbons and chlorofluorocarbons. Supercritical water (T_c = 374°C) is of interest as a substitute for organic solvents to minimize waste in extraction and reaction processes. Additionally, it is used for hydrothermal oxidation of hazardous organic wastes (also called supercritical water oxidation) and hydrothermal synthesis. (See also Sec. 15 for additional discussion of supercritical fluid separation processes.)

PHYSICAL PROPERTIES OF PURE SUPERCRITICAL FLUIDS

Thermodynamic Properties The variation in solvent strength of a SCF from gaslike to liquidlike values (see Table 20-12) may be described qualitatively in terms of the density ρ, as shown in Fig. 20-17, or the solubility parameter.

Similar characteristics are observed for other density-dependent variables including enthalpy, entropy, viscosity, and diffusion coefficient. Above the critical temperature, it is possible to tune the solvent

TABLE 20-10 Advantages of Supercritical Fluid Separations

Solvent strength is adjustable to tailor selectivities and yields.
Diffusion coefficients are higher and viscosities lower, compared with liquids.
Low surface tension favors wetting and penetration of small pores.
There is rapid diffusion of CO_2 through condensed phases, e.g., polymers and ionic liquids.
Solvent recovery is fast and complete, with minimal residue in product.
Collapse of structure due to capillary forces is prevented during solvent removal.
Properties of CO_2 as a solvent:
- Environmentally acceptable solvent for waste minimization, nontoxic, nonflammable, inexpensive, usable at mild temperatures.

Properties of water as a solvent:
- Nontoxic, nonflammable, inexpensive substitute for organic solvents.
- Extremely wide variation in solvent strength with temperature and pressure.

TABLE 20-11 Commercial Applications of Supercritical Fluid Separations Technology

Extraction of foods and pharmaceuticals
- Caffeine from coffee and tea
- Flavors, cholesterol, and fat from foods
- Nicotine from tobacco
- Solvents from pharmaceutical compounds and drugs from natural sources

Extraction of volatile substances from substrates
- Drying and aerogel formation
- Cleaning fabrics, quartz rods for light guide fibers, residues in microelectronics
- Removal of monomers, oligomers, and solvent from polymers

Fractionation
- Residuum oil supercritical extraction (ROSE) (petroleum deasphalting)
- Polymer and edible oils fractionation

CO_2 enhanced oil recovery
Analytical SCF extraction and chromatography
Infusion of materials into polymers (dyes, pharmaceuticals)
Reactive separations
- Extraction of *sec*-butanol from isobutene
- Polymerization to form Teflon
- Depolymerization, e.g., polyethylene terephthalate and cellulose hydrolysis
- Hydrothermal oxidation of organic wastes in water

Crystallization, particle formation, and coatings
- Antisolvent crystallization, rapid expansion from supercritical fluid solution (RESS)
- Particles from gas saturated solutions
- Crystallization by reaction to form metals, semiconductors (e.g., Si), and metal oxides including nanocrystals
- Supercritical fluid deposition

TABLE 20-12 Physical Properties of a Supercritical Fluid Fall between Those of a Typical Gas and Liquid

	Liquid	Supercritical fluid	Gas
Density, g/mL	1	0.05 – 1	10^{-3}
Viscosity, Pa·s	10^{-3}	10^{-4} – 10^{-5}	10^{-5}
Diffusion coefficients, cm^2/s	10^{-5}	10^{-3}	10^{-1}
Surface tension, mN/m	20–50	0	0

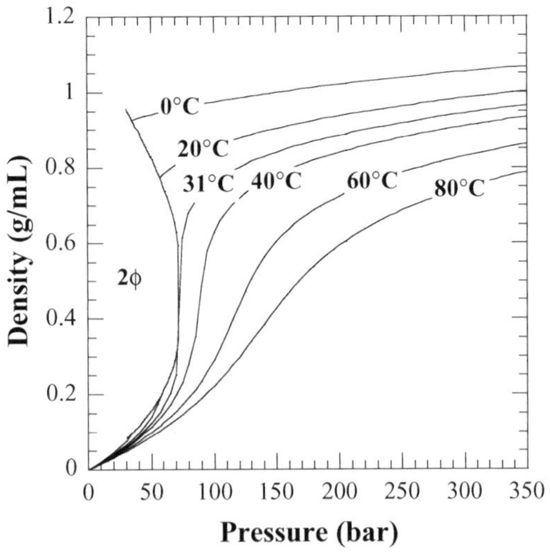

FIG. 20-17 Density versus pressure and temperature for CO_2. (T_c = 31.1°C, P_c = 73.8 bar.)

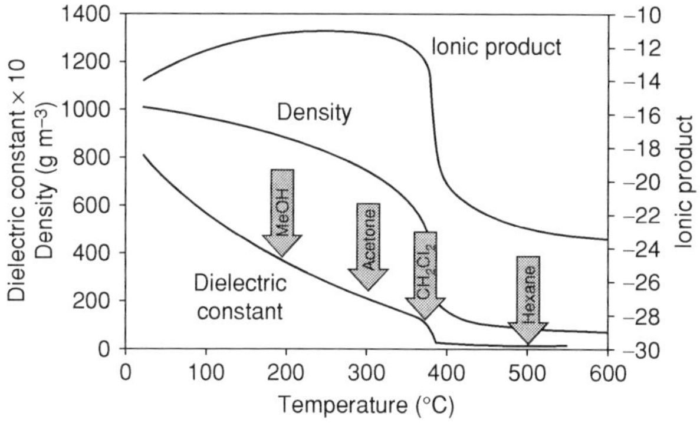

FIG. 20-18 Physical properties of water versus temperature at 240 bar. [*Reprinted from Kritzer and Dinjus, "An Assessment of Supercritical Water Oxidation (SCWO): Existing Problems, Possible Solutions and New Reactor Concepts,"* Chem. Eng. J., *vol.* **83**(3), *pp. 207–214, copyright 2001, with permission form Elsevier.*]

strength continuously over a wide range with either a small isothermal pressure change or a small isobaric temperature change. This unique ability to tune the solvent strength of a SCF may be used to extract and then recover selected products. A good indicator of the van der Waals forces contributed by a SCF is obtained by multiplying ρ by the molecular polarizability α, which is a constant for a given molecule. CO_2's small αρ and low solvent strength are more like those of a fluorocarbon than those of a hydrocarbon.

Water, a key SCF, undergoes profound changes upon heating to the critical point. It expands by a factor of 3, losing about two-thirds of the hydrogen bonds, the dielectric constant drops from 80 to 5 (Shaw et al., op. cit.), and the ionic product falls several orders or magnitude (see Fig. 20-18). At lower densities, supercritical water (SCW) behaves as a "nonaqueous" solvent, and it dissolves many organics and even gases such as O_2. Here it does not solvate ions significantly.

Transport Properties Although the densities of SCFs can approach those of conventional liquids, transport properties are more favorable because viscosities remain lower and diffusion coefficients remain higher. Furthermore, CO_2 diffuses through condensed-liquid phases (e.g., adsorbents and polymers) faster than do typical solvents which have larger molecular sizes. For example, at 35°C the estimated pyrene diffusion coefficient in polymethylmethacrylate increases by 4 orders of magnitude when the CO_2 content is increased from 8 to 17 wt % with pressure [Cao, Johnston, and Webber, *Macromolecules*, **38**(4), 1335–1340 (2005)].

PHASE EQUILIBRIA

Liquid-Fluid Equilibria Nearly all binary liquid-fluid phase diagrams can be conveniently placed in one of six classes (Prausnitz, Lichtenthaler, and de Azevedo, *Molecular Thermodynamics of Fluid Phase Equilibria*, 3d ed., Prentice-Hall, Upper Saddle River, N.J., 1998). Two-phase regions are represented by an area and three-phase regions by a line. In class I, the two components are completely miscible, and a single critical mixture curve connects their critical points. Other classes may include intersections between three phase lines and critical curves. For a ternary system, the slopes of the tie lines (distribution coefficients) and the size of the two-phase region can vary significantly with pressure as well as temperature due to the compressibility of the solvent.

Solid-Fluid Equilibria The solubility of the solid is very sensitive to pressure and temperature in compressible regions, where the solvent's density and solubility parameter are highly variable. In contrast, plots of the log of the solubility versus density at constant temperature often exhibit fairly simple linear behavior (Fig. 20-19). To understand the role of solute-solvent interactions on solubilities and selectivities, it is instructive to define an enhancement factor E as the actual solubility y_2 divided by the solubility in an ideal gas, so that $E = y_2P/P_2^{sat}$, where P_2^{sat} is the vapor pressure. The solubilities in CO_2 are governed primarily by vapor pressures, a property of the solid

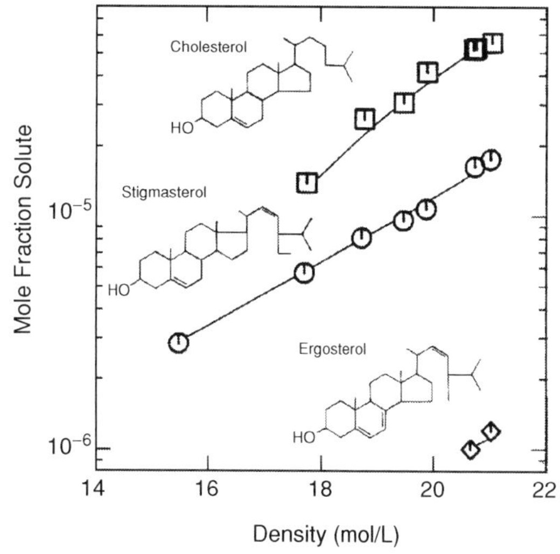

FIG. 20-19 Solubility of sterols in pure CO_2 at 35°C [Wong and Johnston, *Biotech. Prog.*, **2**, 29 (1986)].

crystals, and only secondarily by solute-solvent interactions in the SCF phase. For example, for a given fluid at a particular T and P, the E's are similar for the three sterols, each containing one hydroxyl group, even though the actual solubilities vary by many orders of magnitude (Fig. 20-19).

Polymer-Fluid Equilibria and the Glass Transition Most polymers are insoluble in CO_2, yet CO_2 can be quite soluble in the polymer-rich phase. The solubility in CO_2 may be increased by a combination of lowering the cohesive energy density (which is proportional to the surface tension of the polymer [O'Neill et al., *Ind. Eng. Chem. Res.*, **37,** 3067–79 (1998)]), branching, and the incorporation of either acetate groups in the side chain or carbonate groups in the backbone of the polymer [Sarbu, Styranec, and Beckman, *Nature,* **405,** 165–168 (2000)]. Polyfluoromethacrylates are extremely soluble, and functionalized polyethers and copolymers of cyclic ethers and CO_2 have been shown to be more soluble than most other nonfluorinated polymers. The sorption of CO_2 into silicone rubber is highly dependent upon temperature and pressure, since these properties have a large effect on the density and activity of CO_2. For glassy polymers, sorption isotherms are more complex, and hysteresis between the pressurization and depressurization steps may appear. Furthermore, CO_2 can act as a plasticizer and depress the glass transition temperature by as much as 100°C or even more, producing large changes in mechanical properties and diffusion coefficients. This phenomenon is of interest in conditioning membranes for separations and in commercial foaming of polymers to reduce VOC emissions.

Cosolvents and Complexing Agents Many nonvolatile polar substances cannot be dissolved at moderate temperatures in nonpolar fluids such as CO_2. Cosolvents (also called entrainers) such as alcohols and acetone have been added to fluids to raise the solvent strength for organic solutes and even metals. The addition of only 2 mol % of the complexing agent tri-*n*-butyl phosphate (TBP) to CO_2 increases the solubility of hydroquinone by a factor of 250 due to Lewis acid-base interactions.

Surfactants and Colloids in Supercritical Fluids Because very few nonvolatile molecules are soluble in CO_2, many types of hydrophilic or lipophilic species may be dispersed in the form of polymer latexes (e.g., polystyrene), microemulsions, macroemulsions, and inorganic suspensions of metals and metal oxides (Shah et al., op. cit.). The environmentally benign, nontoxic, and nonflammable fluids water and CO_2 are the two most abundant and inexpensive solvents on earth. Fluorocarbon and hydrocarbon-based surfactants have been used to form reverse micelles, water-in-CO_2

microemulsions (2- to 10-nm droplets) and macroemulsions (50-nm to 5-μm droplets) in SCFs including CO_2. These organized molecular assemblies extend SCF technology to include nonvolatile hydrophilic solutes and ionic species such as amino acids and even proteins. Surfactant micelles or microemulsions are used commercially in dry cleaning and have been proposed for applications including polymerization, formation of inorganic and pharmaceutical particles, and removal of etch/ash residues from low-k dielectrics used in microelectronics. CO_2-in-water macroemulsions, stabilized by surfactants with the proper hydrophilic-CO_2-philic balance, are used in enhanced oil recovery to raise the viscosity of the flowing CO_2 phase for mobility control. Alkane ligands with various head groups have been used to stabilize inorganic nanocrystals in SCW and to stabilize Si and Ge nanocrystals in SCF hydrocarbons and CO_2 at temperatures from 350 to 500°C. Furthermore, colloids may be stabilized by electrostatic stabilization in CO_2 [Smith, Ryoo, and Johnston, *J. Phys. Chem. B.*, **109**(43), 20155 (2005)].

Phase Equilibria Models Two approaches are available for modeling the fugacity of a solute in a SCF solution. The compressed gas approach includes a fugacity coefficient which goes to unity for an ideal gas. The expanded liquid approach is given as

$$f_i^L = x_i \gamma_i(P^\circ, x_i) f_i^{\circ L}(P^\circ) \exp\left(\int_{P^\circ}^{P} \frac{\bar{v}_i}{RT} dP\right) \tag{20-9}$$

where x_i is the mole fraction, γ_i is the activity coefficient, P° and f_i° are the reference pressure and fugacity, respectively, and $\bar{v}_i$ is the partial molar volume of component i. A variety of equations of state have been applied in each approach, ranging from simple cubic equations such as the Peng-Robinson equation of state to the more complex statistical associating fluid theory (SAFT) (Prausnitz et al., op. cit.). SAFT is successful in describing how changes in H bonding of SCF water influence thermodynamic and spectroscopic properties.

MASS TRANSFER

Experimental gas-solid mass-transfer data have been obtained for naphthalene in CO_2 to develop correlations for mass-transfer coefficients [Lim, Holder, and Shah, *Am. Chem. Soc. Symp. Ser.*, **406**, 379 (1989)]. The mass-transfer coefficient increases dramatically near the critical point, goes through a maximum, and then decreases gradually. The strong natural convection at SCF conditions leads to higher mass-transfer rates than in liquid solvents. A comprehensive mass-transfer model has been developed for SCF extraction from an aqueous phase to CO_2 in countercurrent columns [Seibert and Moosberg, *Sep. Sci. Technol.*, **23**, 2049 (1988); Brunner, op. cit.].

PROCESS CONCEPTS IN SUPERCRITICAL FLUID EXTRACTION

Figure 20-20 shows a one-stage extraction process that utilizes the adjustability of the solvent strength with pressure or temperature. The solvent flows through the extraction chamber at a relatively high pressure to extract the components of interest from the feed. The products are then recovered in the separator by depressurization, and the solvent is recompressed and recycled. The products can also be precipitated from the extract phase by raising the temperature after the extraction to lower the solvent density. Multiple extractions or multiple stages may be used with various profiles, e.g., successive increases in pressure or decreases in pressure. Solids may be processed continuously or semicontinuously by pumping slurries or by using lock hoppers. For liquid feeds, multistage separation may be achieved by continuous countercurrent extraction, much as in conventional liquid-liquid extraction. In SCF chromatography, selectivity may be tuned with pressure and temperature programming, with greater numbers of theoretical stages than in liquid chromatography and lower temperatures than in gas chromatography.

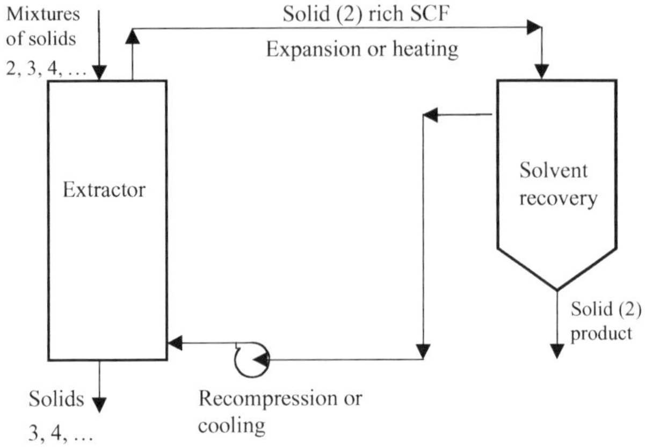

FIG. 20-20 Idealized diagram of a supercritical fluid extraction process for solids.

APPLICATIONS

Decaffeination of Coffee and Tea This application is driven by the environmental acceptability and nontoxicity of CO_2 as well as by the ability to tailor the extraction with the adjustable solvent strength. It has been practiced industrially for more than two decades. Caffeine may be extracted from green coffee beans, and the aroma is developed later by roasting. Various methods have been proposed for recovery of the caffeine, including washing with water and adsorption.

Extraction of Flavors, Fragrances, Nutraceuticals, and Pharmaceuticals Flavors and fragrances extracted by using supercritical CO_2 are significantly different from those extracted by using steam distillation or solvent extraction. The SCF extract can almost be viewed as a new product due to changes in composition associated with the greater amounts of extraction, as shown in Table 20-13. In many instances the flavor or fragrance of the extract obtained with CO_2 is closer to the natural one relative to steam distillation.

Temperature-Controlled Residuum Oil Supercritical Extraction (ROSE) The Kerr-McGee ROSE process has been used worldwide for over two decades to remove asphaltenes from oil. The extraction step uses a liquid solvent that is recovered at supercritical conditions to save energy, as shown in Fig. 20-21. The residuum is contacted with butane or pentane to precipitate the heavy asphaltene fraction. The extract is then passed through a series of heaters, where it goes from the liquid state to a lower-density SCF state. Because the entire process is carried out at conditions near the critical point, a relatively small temperature change is required to produce a fairly large density change. After the light oils have been removed, the solvent is cooled back to the liquid state and recycled.

Polymer Devolatilization, Fractionation, and Plasticization Supercritical fluids may be used to extract solvent, monomers, and oligomers from polymers, including biomaterials. After extraction the pressure is reduced to atmospheric, leaving little residue in the

TABLE 20-13 Comparison of Percent Yields of Flavors and Fragrances from Various Natural Products*

Natural substance	Steam distillation (% yield)	Supercritical CO_2 (% yield)
Ginger	1.5–3.0	4.6
Garlic	0.06–0.4	4.6
Pepper	1.0–2.6	8–18
Rosemary	0.5–1.1	7.5

*Mukhopadhyay, *Natural Extracts Using Supercritical Carbon Dioxide*, CRC Press, Boca Raton, Fla., 2000; Moyler, Extraction of flavours and fragrances with compressed CO_2, in *Extraction of Natural Products Using Near-Critical Solvents*, King and Bott (eds.), Blackie Academic & Professional, London, 1993.

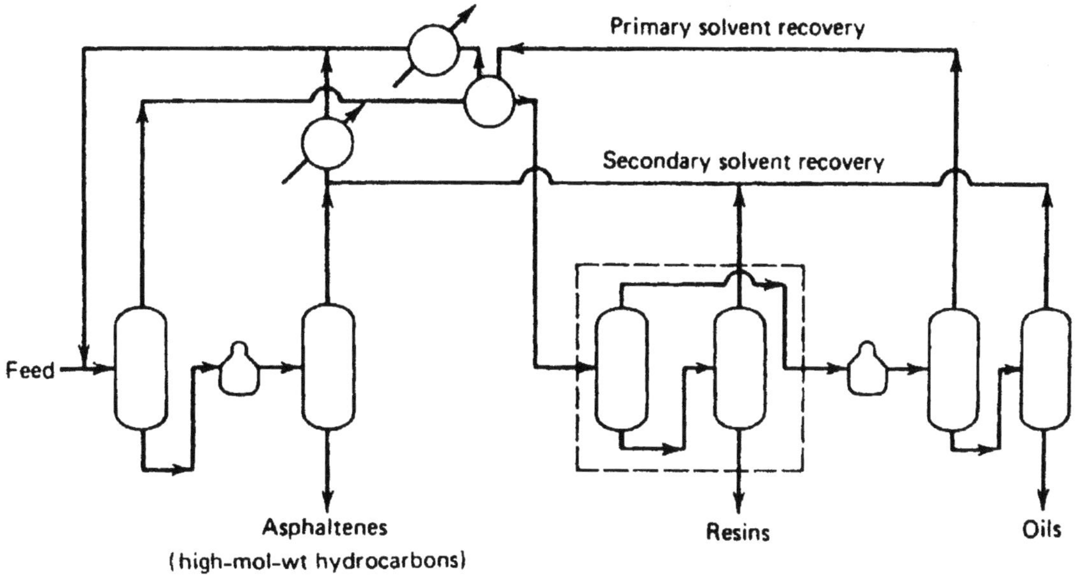

FIG. 20-21 Schematic diagram of the Kerr-McGee ROSE process.

substrate; furthermore, the extracted impurities are easily recovered from the SCF. The swelling and lowering of the glass transition temperature of the polymer by the SCF can increase mass-transfer rates markedly. This approach was used to plasticize block copolymer templates for the infusion of reaction precursors in the synthesis of porous low-*k* dielectrics. For homopolymers, plasticization may be used to infuse dyes, pharmaceuticals, etc., and then the SCF may be removed to trap the solute in the polymer matrix. SCFs may be used to fractionate polymers on the basis of molecular weight and/or composition with various methods for programming pressure and/or temperature (McHugh, op. cit.).

Drying and Aerogel Formation One of the oldest applications of SCF technology, developed in 1932, is SCF drying. The solvent is extracted from a porous solid with a SCF; then the fluid is depressurized. Because the fluid expands from the solid without crossing a liquid-vapor phase transition, capillary forces that would collapse the structure are not present. Using SCF drying, aerogels have been prepared with densities so low that they essentially float in air and look like a cloud of smoke. Also, the process is used in a commercial instrument to dry samples for electron microscopy without perturbing the structure.

Cleaning SCFs such as CO_2 can be used to clean and degrease quartz rods utilized to produce optical fibers, products employed in the fabrication of printed-circuit boards, oily chips from machining operations, precision bearings in military applications, and so on. Research is in progress for removing residues in etch/ash processes in microelectronics.

Microelectronics Processing SCF CO_2 is proposed as a "dry," environmentally benign processing fluid enabling replacement of aqueous and organic solvents in microelectronics fabrication (Desimone, op. cit.). Proposed applications include drying, lithography, solvent spin coating, stripping, cleaning, metal deposition, and chemical mechanical planarization. Due to its low surface tension, tunable solvent strength, and excellent mass-transfer properties, CO_2 offers advantages in wetting of surfaces and small pores, and in removal of contaminants at moderate temperatures.

Precipitation/Crystallization to Produce Nano- and Microparticles Because fluids such as CO_2 are weak solvents for many solutes, they are often effective antisolvents in fractionation and precipitation. In general, a fluid antisolvent may be a compressed gas, a gas-expanded liquid, or a SCF. Typically a liquid solution is sprayed through a nozzle into CO_2 to precipitate a solute. As CO_2 mixes with the liquid phase, it decreases the cohesive energy density (solvent strength) substantially, leading to precipitation of dissolved solutes (e.g., crystals of progesterone). The high diffusion rates of the organic solvent into CO_2 and vice versa can lead to rapid phase separation, and the supersaturation curve may be manipulated to vary the crystalline morphology (Yeo and Kiran, op. cit.).

Nanoparticles of controllable size can be obtained in the supercritical antisolvent-enhanced mass-transfer (SAS-EM) process, which can produce commercial quantities of pharmaceuticals (see Fig. 20-22) [Chattopadhyay and Gupta, *Ind. Engr. Chem. Res.*, **40,** 3530–3539 (2001)]. Here, the solution jet is injected onto an ultrasonic vibrating surface H inside the antisolvent chamber to aid droplet atomization. The particle size is controlled by varying the vibration intensity. For most pharmaceuticals, organic compounds, proteins, and polymers, average particle diameters range from 100 to 1000 nm; even smaller particles may be obtained for certain inorganic compounds.

Rapid Expansion from Supercritical Solution and Particles from Gas Saturated Solutions Rapid expansion from supercritical solution (RESS) of soluble materials may be used to form microparticles or microfibers. A variety of inorganic crystals have been formed naturally and synthetically in SCF water, and organic crystals have been formed in SCF CO_2. Recently, the addition of a solid cosolvent (e.g., menthol, which can be removed later by sublimation) has overcome key limitations by greatly enhancing solubilities in CO_2 and producing smaller nanoparticles by reducing particle-particle coagulation [Thakur and Gupta, *J. Sup. Fluids*, **37,** 307–315 (2006)]. Another approach is to expand the solutions into aqueous solutions containing soluble surfactants to arrest growth due to particle collisions. RESS typically uses dilute solutions. For concentrated solutions, the process is typically referred to as particle formation from gas saturated solutions (PGSS). Here CO_2 lowers the viscosity of the melt to facilitate flow. Union Carbide developed the commercial UNICARB process to replace organic solvents with CO_2 as a diluent in coating applications to reduce volatile organic carbon emissions and form superior coatings. For aqueous solutions, the expansion of CO_2 facilitates atomization, and the resulting cooling may be used to control the freezing of the solute.

Reactive Separations Reactions may be integrated with SCF separation processes to achieve a large degree of control for producing a highly purified product. Reaction products may be recovered by

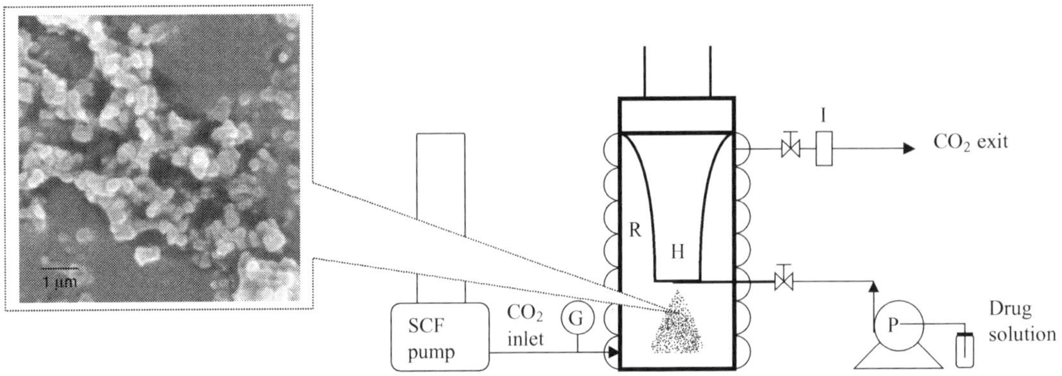

FIG. 20-22 Schematic of supercritical antisolvent with enhanced mass-transfer process to produce nanoparticles of controllable size. R, precipitation chamber; SCF pump, supply of supercritical CO_2; I, inline filter; H, ultrasonic horn; P, pump for drug solution; G, pressure gauge.

volatilization into, or precipitation from, a SCF phase. A classic example is the high-pressure production of polyethylene in SCF ethylene. The molecular weight distribution may be controlled by choosing the temperature and pressure for precipitating the polymer from the SCF phase. Over a decade ago, Idemitsu commercialized a 5000 metric ton per/year (t/yr) integrated reaction and separation process in SCF isobutene. The reaction of isobutene and water produces *sec*-butanol, which is extracted from water by the SCF solvent. SCF solvents have been tested for reactive extractions of liquid and gaseous fuels from heavy oils, coal, oil shale, and biomass. In some cases the solvent participates in the reaction, as in the hydrolysis of coal and heavy oils with SCW. Related applications include conversion of cellulose to glucose in water, delignification of wood with ammonia, and liquefaction of lignin in water.

Gas-expanded liquids (GXLs) are emerging solvents for environmentally benign reactive separation (Eckert et al., op. cit.). GXLs, obtained by mixing supercritical CO_2 with normal liquids, show intermediate properties between normal liquids and SCFs both in solvation power and in transport properties; and these properties are highly tunable by simple pressure variations. Applications include chemical reactions with improved transport, catalyst recycling, and product separation.

Hydrothermal oxidation (HO) [also called supercritical water oxidation (SCWO)] is a reactive process to convert aqueous wastes to water, CO_2, O_2, nitrogen, salts, and other by-products. It is an enclosed and complete water treatment process, making it more desirable to the public than incineration. Oxidation is rapid and efficient in this one-phase solution, so that wastewater containing 1 to 20 wt % organics may be oxidized rapidly in SCW with the potential for higher energy efficiency and less air pollution than in conventional incineration. Temperatures range from about 375 to 650°C and pressures from 3000 to about 5000 psia.

Crystallization by Chemical Reaction

Supercritical Fluid Deposition (SFD) Metal films may be grown from precursors that are soluble in CO_2. The SFD process yields copper films with fewer defects than those possible by using chemical vapor deposition, because increased precursor solubility removes mass-transfer limitations and low surface tension favors penetration of high-aspect-ratio features [Blackburn et al., *Science*, **294**, 141–145 (2001)].

High-Temperature Crystallization The size-tunable optical and electronic properties of semiconductor nanocrystals are attractive for a variety of optoelectronic applications. In solution-phase crystallization, precursors undergo chemical reaction to form nuclei, and particle growth is arrested with capping ligands that passivate the surface. However, temperatures above 350°C are typically needed to crystallize the group IV elements silicon and germanium, due to the covalent network structure. Whereas liquid solvents boil away at these elevated temperatures, SCFs under pressure are capable of solvating the capping ligands to stabilize the nanocrystals (Shah et al., op. cit.). Crystalline Si and Ge nanocrystals, with an average size of 2 to 70 nm, may be synthesized in supercritical CO_2, hexane, or octanol at 400 to 550°C and 20 MPa in a simple continuous flow reactor. UV-visible absorbance and photoluminescence (PL) spectra of Ge nanocrystals of 3- to 4-nm diameter exhibit optical absorbance and PL spectra blue-shifted by approximately 1.7 eV relative to the band gap of bulk Ge, as shown in Fig. 20-23. One-dimensional silicon nanowires may be grown from relatively monodisperse gold nanocrystals stabilized with dodecanethiol ligands, as shown in Fig. 20-24. The first crystalline silicon nanowires with diameters smaller than 5 nm and lengths greater than 1 μm made by any technique were produced in SCF hexane. Hydrothermal crystallization has also been used to produce metal oxide nano- and microcrystals by rapid generation of supersaturation during hydrolysis of precursors, such as metal nitrates, during rapid heating of aqueous solutions.

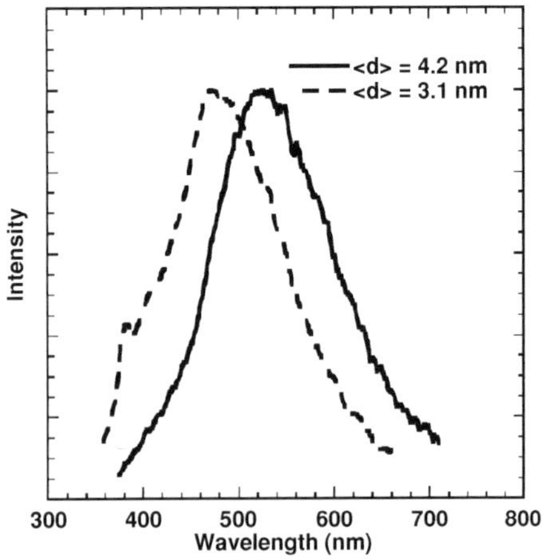

FIG. 20-23 Normalized photoluminescence spectra of 3.1-nm ($\lambda_{excitation}$ = 320 nm) and 4.2-nm ($\lambda_{excitation}$ = 340 nm) Ge nanoparticles dispersed in chloroform at 25°C with quantum yields of 6.6 and 4.6 percent, respectively. [*Reprinted with permission from Lu et al.,* Nano Lett., *4(5), 969–974 (2004). Copyright 2004 American Chemical Society.*]

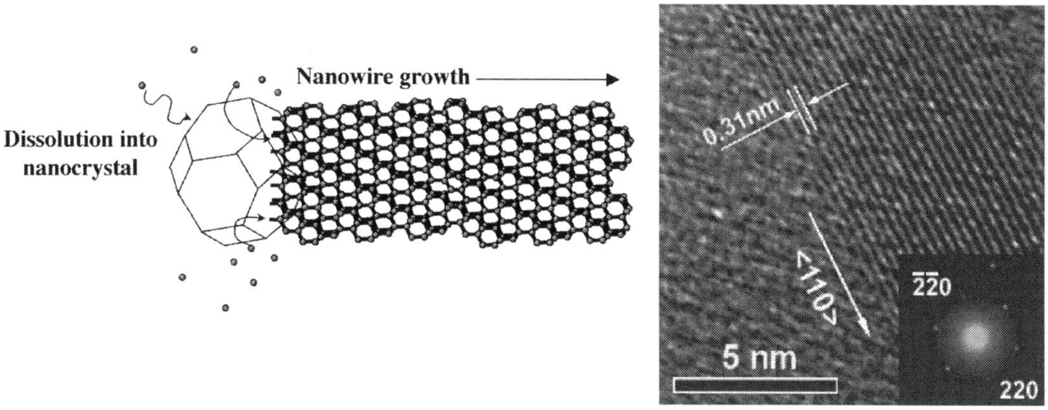

FIG. 20-24 High-resolution TEM image of Si nanowires produced at 500°C and 24.1 MPa in supercritical hexane from gold seed crystals. *Inset:* Electron diffraction pattern indexed for the <111> zone axis of Si indicates <110> growth direction. [*Reprinted with permission from Lu et al.*, Nano Lett., *3(1), 93–99 (2003). Copyright 2003 American Chemical Society.*]

ALTERNATIVE SOLID/LIQUID SEPARATIONS

SEPARATION PROCESSES BASED PRIMARILY ON ACTION IN AN ELECTRIC FIELD

Differences in mobilities of ions, molecules, or particles in an electric field can be exploited to perform useful separations. Primary emphasis is placed on electrophoresis and dielectrophoresis. Analogous separation processes involving magnetic and centrifugal force fields are widely applied in the process industry (see Secs. 18 and 19).

Theory of Electrical Separations

General References: Newman, *Adv. Electrochem. Electrochem. Eng.*, **5,** 87 (1967); *Ind. Eng. Chem.*, **60**(4), 12 (1968). Ptasinski and Kerkhof, *Sep. Sci. Technol.*, **27,** 995 (1992).

For electrolytic solutions, migration of charged species in an electric field constitutes an additional mechanism of mass transfer. Thus the flux of an ionic species N_i in (g·mol)/(cm^2·s) in dilute solutions can be expressed as

$$N_i = -z_i u_i \mathscr{F} c_i \nabla E - D_i \nabla c_i + c_i v \tag{20-10}$$

The ionic mobility u_i is the average velocity imparted to the species under the action of a unit force (per mole). v is the stream velocity, cm/s. In the present case, the electrical force is given by the product of the electric field ∇E in V/cm and the charge $z_i\mathscr{F}$ per mole, where $\mathscr{F}$ is the Faraday constant in C/g equivalent and z_i is the valence of the ith species. Multiplication of this force by the mobility and the concentration c_i [(g·mol)/cm^3] yields the contribution of migration to the flux of the ith species.

The diffusive and convective terms in Eq. (20-10) are the same as in nonelectrolytic mass transfer. The ionic mobility u_i, (g·mol·cm^2)/(J·s), can be related to the ionic-diffusion coefficient D_i, cm^2/s, and the ionic conductance of the ith species λ_i, cm^2/(Ω·g equivalent):

$$u_i = D_i/RT = \lambda_i / |z_i| \mathscr{F}^2 \tag{20-11}$$

where T is the absolute temperature, K; and R is the gas constant, 8.3143 J/(K·mol). Ionic conductances are tabulated in the literature (Robinson and Stokes, *Electrolyte Solutions,* Academic, New York, 1959). For practical purposes, a bulk electrolytic solution is electrically neutral.

$$\sum_i z_i c_i = 0 \tag{20-12}$$

since the forces required to effect an appreciable separation of charge are prohibitively large.

The **current density** (A/cm^2) produced by movement of charged species is described by summing the terms in Eq. (20-13) for all species:

$$i = \mathscr{F} \sum_i z_i \mathbf{N}_i = -\kappa \nabla E - \mathscr{F} \sum_i z_i D_i \nabla c_i \tag{20-13}$$

where the electrical conductivity κ in S/cm is given by

$$\kappa = \mathscr{F}^2 \sum_i z_i^2 u_i c_i \tag{20-14}$$

In solutions of uniform composition, the diffusional terms vanish and Eq. (20-13) reduces to Ohm's law.

Conservation of each species is expressed by the relation

$$\partial c_i / \partial t = -\nabla \cdot \mathbf{N}_i \tag{20-15}$$

provided that the species is not produced or consumed in homogeneous chemical reactions. In two important cases, this conservation law reduces to the equation of convective diffusion:

$$(\partial c_i / \partial t) + \mathbf{v}\nabla \cdot c_i = D \nabla^2 c_i \tag{20-16}$$

First, when a large excess of inert electrolyte is present, the electric field will be small and migration can be neglected for minor ionic components; Eq. (20-16) then applies to these minor components, where D is the ionic-diffusion coefficient. Second, Eq. (20-16) applies when the solution contains only one cationic and one anionic species. The electric field can be eliminated by means of the electroneutrality relation.

In the latter case the diffusion coefficient D of the electrolyte is given by

$$D = (z_+ u_+ D_- - z_- u_- D_+)/(z_+ u_+ - z_- u_-) \tag{20-17}$$

which represents a compromise between the diffusion coefficients of the two ions. When Eq. (20-16) applies, many solutions can be obtained by analogy with heat transfer and nonelectrolytic mass transfer.

Because the solution is electrically neutral, conservation of charge is expressed by differentiating Eq. (20-13):

$$\nabla \cdot \mathbf{i} = 0 = -\kappa \nabla^2 E - \mathscr{F} \sum_i z_i D_i \nabla^2 c_i \tag{20-18}$$

For solutions of uniform composition, Eq. (20-18) reduces to Laplace's equation for the potential:

$$\nabla^2 E = 0 \tag{20-19}$$

This equation is the starting point for determination of the current-density distributions in many electrochemical cells.

Near an interface or at solution junctions, the solution departs from electroneutrality. Charges of one sign may be preferentially adsorbed at the interface, or the interface may be charged. In either case, the charge at the interface is counterbalanced by an equal and opposite charge composed of ions in the solution. Thermal motion prevents this countercharge from lying immediately adjacent to the interface, and the result is a "diffuse-charge layer" whose thickness is on the order of 10 to 100 Å.

A tangential electric field ∇E_t acting on these charges produces a relative motion between the interface and the solution just outside the diffuse layer. In view of the thinness of the diffuse layer, a balance of the tangential viscous and electrical forces can be written

$$\mu(\partial^2 v_t/\partial y^2) + \rho_e \nabla E_t = 0 \tag{20-20}$$

where μ is the viscosity and ρ_e is the electric-charge density, C/cm^3. Furthermore, the variation of potential with the normal distance satisfies Poisson's equation:

$$\partial^2 E/\partial y^2 = -(\rho_e/\varepsilon) \tag{20-21}$$

with ε defined as the **permittivity** of the solution. [The relative dielectric constant is $\varepsilon/\varepsilon_0$, where ε_0 is the permittivity of free space; $\varepsilon_0 = 8.8542 \times 10^{-14}$ C/(V·cm).] Elimination of the electric-charge density between Eqs. (20-20) and (20-21) with two integrations, gives a relation between ∇E_t and the velocity v_0 of the bulk solution relative to the interface.

$$\mu[v_t(\infty) - v_t(0)] = \varepsilon \nabla E_t [E(\infty) - E(0)] \tag{20-22}$$

or

$$v_0 = -(\varepsilon \nabla E_t \zeta/\mu) \tag{20-23}$$

The potential difference across the mobile part of the diffuse-charge layer is frequently called the **zeta potential,** $\zeta = E(0) - E(\infty)$. Its value depends on the composition of the electrolytic solution as well as on the nature of the particle-liquid interface.

There are four related electrokinetic phenomena which are generally defined as follows: *electrophoresis*—the movement of a charged surface (i.e., suspended particle) relative to a stationary liquid induced by an applied electrical field, *sedimentation potential*—the electric field which is crested when charged particles move relative to a stationary liquid, *electroosmosis*—the movement of a liquid relative to a stationary charged surface (i.e., capillary wall), and *streaming potential*—the electric field which is created when liquid is made to flow relative to a stationary charged surface. The effects summarized by Eq. (20-23) form the basis of these electrokinetic phenomena.

For many particles, the diffuse-charge layer can be characterized adequately by the value of the zeta potential. For a spherical particle of radius r_0 which is large compared with the thickness of the diffuse-charge layer, an electric field uniform at a distance from the particle will produce a tangential electric field which varies with position on the particle. Laplace's equation [Eq. (20-19)] governs the distribution of potential outside the diffuse-charge layer; also, the Navier-Stokes equation for a creeping-flow regime can be applied to the velocity distribution. On account of the thinness of the diffuse-charge layer, Eq. (20-23) can be used as a local boundary condition, accounting for the effect of this charge in leading to movement of the particle relative to the solution. The result of this computation gives the velocity of the particle as

$$v = \varepsilon\zeta \nabla E/\mu \tag{20-24}$$

and it may be convenient to tabulate the mobility of the particle

$$U = v/\nabla E = \varepsilon\zeta/\mu \tag{20-25}$$

rather than its zeta potential. Note that this mobility gives the velocity of the particle for unit electric field rather than for unit force on the particle. Related equations can be developed for the velocity of electroosmotic flow. The subsections presented below ("Electrophoresis," "Electrofiltration," and "Cross-Flow–Electrofiltration") represent both established and emerging commercial applications of electrokinetic phenomena.

Electrophoresis

GENERAL REFERENCE: Wankat, *Rate-Controlled Separations,* Elsevier, London, 1990.

Electrophoretic Mobility Macromolecules move at speeds measured in tenths of micrometers per second in a field (gradient) of 1 V/cm. Larger particles such as bubbles or bacteria move up to 10 times as fast because U is usually higher. To achieve useful separations, therefore, voltage gradients of 10 to 100 V/cm are required. High voltage gradients are achieved only at the expense of power dissipation within the fluid, and the resulting heat tends to cause undesirable convection currents.

Mobility is affected by the dielectric constant and viscosity of the suspending fluid, as indicated in Eq. (20-25). The ionic strength of the fluid has a strong effect on the thickness of the double layer and hence on ζ. As a rule, mobility varies inversely as the square root of ionic strength [Overbeek, *Adv. Colloid Sci.,* **3,** 97 (1950)].

Modes of Operation There is a close analogy between sedimentation of particles or macromolecules in a gravitational field and their electrophoretic movement in an electric field. Both types of separation have proved valuable not only for analysis of colloids but also for preparative work, at least in the laboratory. Electrophoresis is applicable also for separating mixtures of simple cations or anions in certain cases in which other separating methods are ineffectual.

Electrodecantation or electroconvection is one of several operations in which one mobile component (or several) is to be separated out from less mobile or immobile ones. The mixture is introduced between two vertical semipermeable membranes; for separating cations, anion membranes are used, and vice versa. When an electric field is applied, the charged component migrates to one or another of the membranes; but since it cannot penetrate the membrane, it accumulates at the surface to form a dense concentrated layer of particles which will sink toward the bottom of the apparatus. Near the top of the apparatus immobile components will be relatively pure. Murphy [*J. Electrochem. Soc.,* **97**(11), 405 (1950)] has used silver-silver chloride electrodes in place of membranes. Frilette [*J. Phys. Chem.,* **61,** 168 (1957)], using anion membranes, partially separated H^+ and Na^+, K^+ and Li^+, and K^+ and Na^+.

Countercurrent electrophoresis can be used to split a mixture of mobile species into two fractions by the electrical analog of elutriation. In such countercurrent electrophoresis, sometimes termed an ion still, a flow of the suspending fluid is maintained parallel to the direction of the voltage gradient. Species which do not migrate fast enough in the applied electric field will be physically swept out of the apparatus. An apparatus based mainly on this principle but using also natural convection currents has been developed (Bier, *Electrophoresis,* vol. II, Academic, New York, 1967).

Membrane electrophoresis which is based upon differences in ion mobility, has been studied by Glueckauf and Kitt [*J. Appl. Chem.,* **6,** 511 (1956)]. Partial exclusion of coions by membranes results in large differences in coion mobilities. Superposing a cation and an anion membrane gives high transference numbers (about 0.5) for both cations and anions while retaining the selectivity of mobilities. Large voltages are required, and flow rates are low.

In continuous-flow zone electrophoresis the "solute" mixture to be separated is injected continuously as a narrow source within a body of carrier fluid flowing between two electrodes. As the "solute" mixture passes through the transverse field, individual components migrate sideways to produce zones which can then be taken off separately downstream as purified fractions.

Resolution depends upon differences in mobilities of the species. Background electrolyte of low ionic strength is advantageous, not only to increase electrophoretic (solute) mobilities, but also to achieve low electrical conductivity and thereby to reduce the thermal-convection current for any given field [Finn, in Schoen (ed.), *New Chemical Engineering Separation Techniques,* Interscience, New York, 1962].

The need to limit the maximum temperature rise has resulted in two main types of apparatus, illustrated in Fig. 20-25. The first consists of multicomponent ribbon separation units—apparatus capable of separating small quantities of mixtures which may contain few or many species. In general, such units operate with high voltages, low

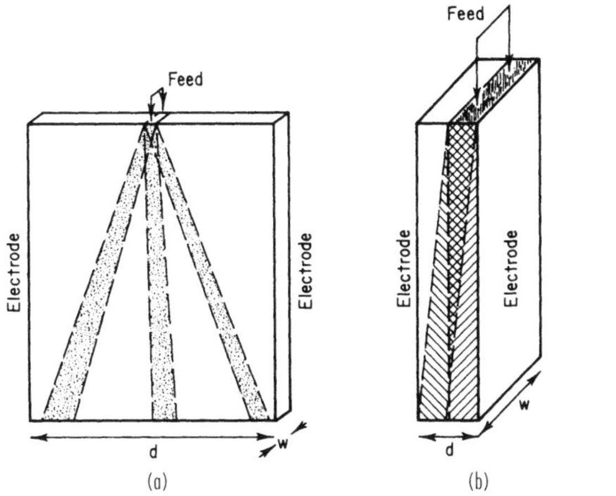

FIG. 20-25 Types of arrangement for zone electrophoresis or electrochromatography. (*a*) Ribbon unit, with $d > w$; cooling at side faces. (*b*) Block unit, with $w > d$; cooling at electrodes.

currents, a large transverse dimension, and a narrow thickness between cooling faces. Numerous units developed for analytic chemistry, generally with filter-paper curtains but sometimes with granular "anticonvectant" packing, are of this type. The second type consists of block separation units—apparatus designed to separate larger quantities of a mixture into two (or at most three) species or fractions. Such units generally use low to moderate voltages and high currents, with cooling by circulation of cold electrolyte through the electrode compartments. Scale-up can readily be accomplished by extending the thickness dimension *w*.

Both types of units have generally been operated in trace mode; that is, "background" or "elutant" electrolyte is fed to the unit along with the mixture to be separated. A desirable and possible means of operation for preparative applications is in bulk mode, in which one separated component follows the other without background electrolyte being present, except that other ions may be required to bracket the separated zones. Overlap regions between components should be recycled, and pure components collected as products.

For block units, the need to stabilize flow has given rise to a number of distinct techniques.

Free flow. Dobry and Finn [*Chem. Eng. Prog.*, **54,** 59 (1958)] used upward flow, stabilized by adding methyl cellulose, polyvinyl alcohol, or dextran to the background solution. Upward flow was also used in the electrode compartments, with cooling efficiency sufficient to keep the main solution within 1°C of entering temperature.

Density gradients to stabilize flow have been employed by Philpot [*Trans. Faraday Soc.*, **36,** 38 (1940)] and Mel [*J. Phys. Chem.*, **31,** 559 (1959)]. Mel's Staflo apparatus [*J. Phys. Chem.*, **31,** 559 (1959)] has liquid flow in the horizontal direction, with layers of increasing density downward produced by sucrose concentrations increasing to 7.5 percent. The solute mixture to be separated is introduced in one such layer. Operation at low electrolyte concentrations, low voltage gradients, and low flow rates presents no cooling problem.

Packed beds. A packed cylindrical electrochromatograph 9 in (23 cm) in diameter and 48 in (1.2 m) high, with operating voltages in the 25- to 100-V range, has been developed by Hybarger, Vermeulen, and coworkers [*Ind. Eng. Chem. Process Des. Dev.*, **10,** 91 (1971)]. The annular bed is separated from inner and outer electrodes by porous ceramic diaphragms. The unit is cooled by rapid circulation of cooled electrolyte between the diaphragms and the electrodes.

An interesting modification of zone electrophoresis resolves mixtures of ampholytes on the basis of **differing isoelectric points** rather than differing mobilities. Such **isoelectric spectra** develop when a pH gradient is established parallel to the electric field. Each species then migrates until it arrives at the region of pH where it possesses no net surface charge. A strong focusing effect is thereby achieved [Kolin, in Glick (ed.), *Methods of Biochemical Analysis*, vol. VI, Interscience, New York, 1958].

Electrofiltration

GENERAL REFERENCE: P. Krishnaswamy and P. Klinkowski, "Electrokinetics and Electrofiltration," in *Advances in Solid-Liquid Separation*, H. S. Muralidhara (ed.), Battelle Press, Columbus, OH, 1986.

Process Concept The application of a direct electric field of appropriate polarity when filtering should cause a net charged-particle migration relative to the filter medium (electrophoresis). The same direct electric field can also be used to cause a net fluid flow relative to the pores in a fixed filter cake or filter medium (electroosmosis). The exploitation of one or both of these phenomena form the basis of conventional electrofiltration.

In conventional filtration, often the object is to form a high-solids-content filter cake. At a single-filter surface, a uniform electric field can be exploited in one of two ways. The first method of exploitation occurs when the electric field is of a polarity such that the charged-particle migration occurs toward the filter medium. In this case, the application of the electric field increases the velocity of the solid particles toward the filter surface (electrosedimentation), thereby hastening the clarification of the feed suspension and, at the same time, increasing the compaction of the filter cake collected on the filter surface. In this first case, electroosmotic flow occurs in a direction away from the filter media. The magnitude of the pressure-driven fluid flow toward the filter surface far exceeds the magnitude of the electroosmotic flow away from the surface so that the electroosmotic flow results in only a minor reduction of the rate of production of filtrate. The primary benefits of the applied electric field in this case are increased compaction, and hence increased dewatering, of the filter cake and an increased rate of sedimentation or movement of the particles in bulk suspension toward the filter surface.

The second method of exploitation occurs when the electric field is of a polarity such that the charged-particle migration occurs away from the filter medium. The contribution to the net-particle velocity of the electrophoretically induced flow away from the filter medium is generally orders of magnitude less than the contribution to the net-particle velocity of the flow induced by drag due to the pressure-induced flow of the bulk liquid toward the filter media. (In conventional or cake filtration, the velocity of liquid in dead-end flow toward the filter is almost always sufficient to overcome any electrophoretic migration of particles away from the filter media so that the prevention of the formation of filter cake is not an option. This will not necessarily be the case for cross-flow electrofiltration.) The primary enhancement to filtration caused by the application of an electrical field in this manner is the increase in the filtrate flux due to electroosmotic flow through the filter cake. This electroosmotic flow is especially beneficial during the latter stages of filtration when the final filter-cake thickness has been achieved. At this stage, electroosmosis can be exploited to draw filtrate out from the pore structure of the filter cake. This type of drying of the filter cake is sometimes called **electroosmotic dewatering.**

Commercial Applications Krishnaswamy and Klinkowski, op. cit., describe the Dorr-Oliver EAVF®. The EAVF® combines vacuum filtration with electrophoresis and electroosmosis and has been described as a series of parallel platelike electrode assemblies suspended in a tank containing the slurry to be separated. When using the EAVF®, solids are collected at both electrodes, one collecting a compacted cake simply by electrophoretic attraction and the second collecting a compacted cake though vacuum filtration coupled with electroosmotic dewatering. Upon the completion of a collection cycle, the entire electrode assembly is withdrawn from the slurry bath and the cake is removed. The EAVF® is quoted as being best suited for the dewatering of ultrafine slurries (particle sizes typically less than 10 μm).

Cross-Flow–Electrofiltration

GENERAL REFERENCES: Henry, Lawler, and Kuo, *Am. Inst. Chem. Eng. J.*, **23**(6), 851 (1977). Kuo, Ph.D. dissertation, West Virginia University, 1978.

Process Concept The application of a direct electric field of appropriate polarity when filtering should cause a net charged-particle migration away from the filter medium. This electrophoretic migration will prevent filter-cake formation and the subsequent reduction of filter performance. An additional benefit derived from the imposed electric field is an electroosmotic flux. The presence of this flux in the membrane and in any particulate accumulation may further enhance the filtration rate.

Cross-flow–electrofiltration (CF-EF) is the multifunctional separation process which combines the electrophoretic migration present in electrofiltration with the particle diffusion and radial-migration forces present in cross-flow filtration (CFF) (microfiltration includes cross-flow filtration as one mode of operation in "Membrane Separation Processes" which appears later in this section) in order to reduce further the formation of filter cake. Cross-flow–electrofiltration can even eliminate the formation of filter cake entirely. This process should find application in the filtration of suspensions when there are charged particles as well as a relatively low conductivity in the continuous phase. Low conductivity in the continuous phase is necessary in order to minimize the amount of electrical power necessary to sustain the electric field. Low-ionic-strength aqueous media and nonaqueous suspending media fulfill this requirement.

Cross-flow–electrofiltration has been investigated for both aqueous and nonaqueous suspending media by using both rectangular- and tubular-channel processing configurations (Fig. 20-26). Henry, Lawler, and Kuo (op. cit.), using a rectangular-channel system with a 0.6-μ-pore-size polycarbonate Nuclepore filtration membrane, investigated CF-EF for 2.5-μm kaolin-water and 0.5- to 2-μm oil-in-water emulsion systems. Kuo (op. cit.), using similar equipment, studied 5-μm kaolin-water, ~100-μm Cr_2O_3-water, and ~6-μm Al_2O_3-methanol and/or -butanol systems. For both studies electrical fields of 0 to 60 V/cm were used for aqueous systems, and to 5000 V/cm were used for nonaqueous systems. The studies covered a wide range of processing variables in order to gain a better understanding of CF-EF fundamentals. Lee, Gidaspow, and Wasan [*Ind. Eng. Chem. Fundam.*, **19**(2), 166 (1980)] studied CF-EF by using a porous stainless-steel tube (pore size = 5 μm) as the filtration medium. A platinum wire running down the center of the tube acted as one electrode, while the porous steel tube itself acted as the other electrode. Nonaqueous suspensions of 0.3- to 2-μm Al_2O_3-tetralin and a coal-derived liquid diluted with xylene and tetralin were studied. By operating with applied electric fields (1000 to 10,000 V/cm) above the critical voltage, clear particle-free filtrates were produced. It should be noted that the pore size of the stainless-steel filter medium (5 μm) was greater than the particle size of the suspended Al_2O_3 solids (0.3 to 2 μm). Cross-flow–electrofiltration has also been applied to biological systems. Brors, Kroner, and Deckwer [ECB6: Proc. 6th Eur. Cong. Biotech., 511 (1994)] separated malate dehydrogenase from the cellular debris of Baker's yeast using CF-EF. A two- to fivefold increase in the specific enzyme transport rate was reported when electric field strengths of 20 to 40 V/cm were used.

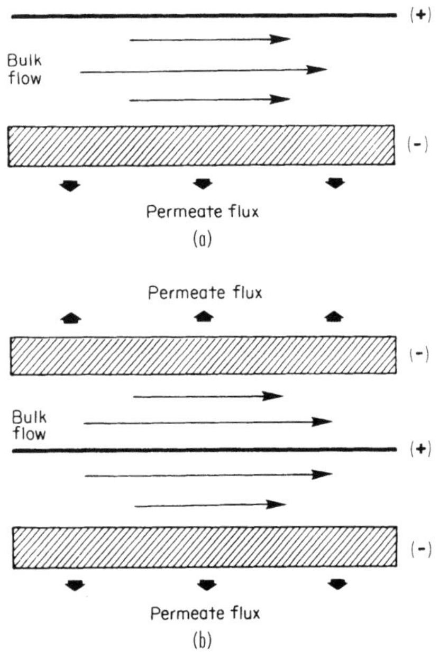

FIG. 20-26 Alternative electrode configurations for cross-flow–electrofiltration.

Theory Cross-flow–electrofiltration can theoretically be treated as if it were cross-flow filtration with superimposed electrical effects. These electrical effects include electroosmosis in the filter medium and cake and electrophoresis of the particles in the slurry. The addition of the applied electric field can, however, result in some qualitative differences in permeate-flux-parameter dependences.

The membrane resistance for CF-EF can be defined by specifying two permeate fluxes as

$$J_{om} = \Delta P/R_{om} \qquad (20\text{-}26)$$

$$J_m = \Delta P/R_m \qquad (20\text{-}27)$$

where J_{om} is the flux through the membrane in the absence of an electric field and any other resistance, m/s; J_m is the same flux in the presence of an electric field; and R_{om} is the membrane resistance in the absence of an electric field, (N·s)/m^3. When electroosmotic effects do occur,

$$J_m = J_{om} + K_m E \qquad (20\text{-}28)$$

where K_m is the electroosmotic coefficient of the membrane, m^2/(V·s); and E is the applied-electric-field strength, V/m. Equations (20-26), (20-27), and (20-28) can be combined and rearranged to give Eq. (20-29), the membrane resistance in the presence of an electric field.

$$R_m = \frac{R_{om}}{1 + \left(\dfrac{K_m E}{J_{om}}\right)} \qquad (20\text{-}29)$$

Similarly, cake resistance can be represented as

$$R_c = \frac{R_{oc}}{1 + \left(\dfrac{K_c E}{J_{oc}}\right)} \qquad (20\text{-}30)$$

where J_{oc} is the flux through the cake in the absence of an electric field or any other resistance, R_{oc} is the cake resistance in the absence of an electric field, and K_c is the electroosmotic coefficient of the cake. The cake resistance is not a constant but is dependent upon the cake thickness, which is in turn a function of the transmembrane pressure drop and electrical-field strength.

Particulate systems require the addition of the term $\mu_e E$ in order to account for the electrophoretic migration of the particle. The constant μ_e is the electrophoretic mobility of the particle, m^2/(V·s). For the case of the CF-EF, the film resistance R_f can be represented as

$$R_f = \frac{\Delta P}{k \ln\left(\dfrac{C_s}{C_b}\right) + U_r + \mu_e E} \qquad (20\text{-}31)$$

The resistances, when incorporated into equations descriptive of cross-flow filtration, yield the general expression for the permeate flux for particulate suspensions in cross-flow–electrofiltration systems.

There are three distinct regimes of operation in CF-EF. These regimes (Fig. 20-27) are defined by the magnitude of the applied electric field with respect to the critical voltage E_c. The critical voltage is defined as the voltage at which the net particle migration velocity toward the filtration medium is zero. At the critical voltage, there is a balance between the electrical-migration and radial-migration velocities away from the filter and the velocity at which the particles are swept toward the filter by bulk flow. There is no diffusive transport at $E = E_c$ (Fig. 20-27*b*) because there is no gradient in the particle concentration normal to the filter surface. At field strengths below the critical voltage (Fig. 20-27*a*), all migration velocities occur in the same direction as in the cross-flow-filtration systems discussed earlier. At values of applied voltage above the critical voltage (Fig. 20-27*c*) qualitative differences

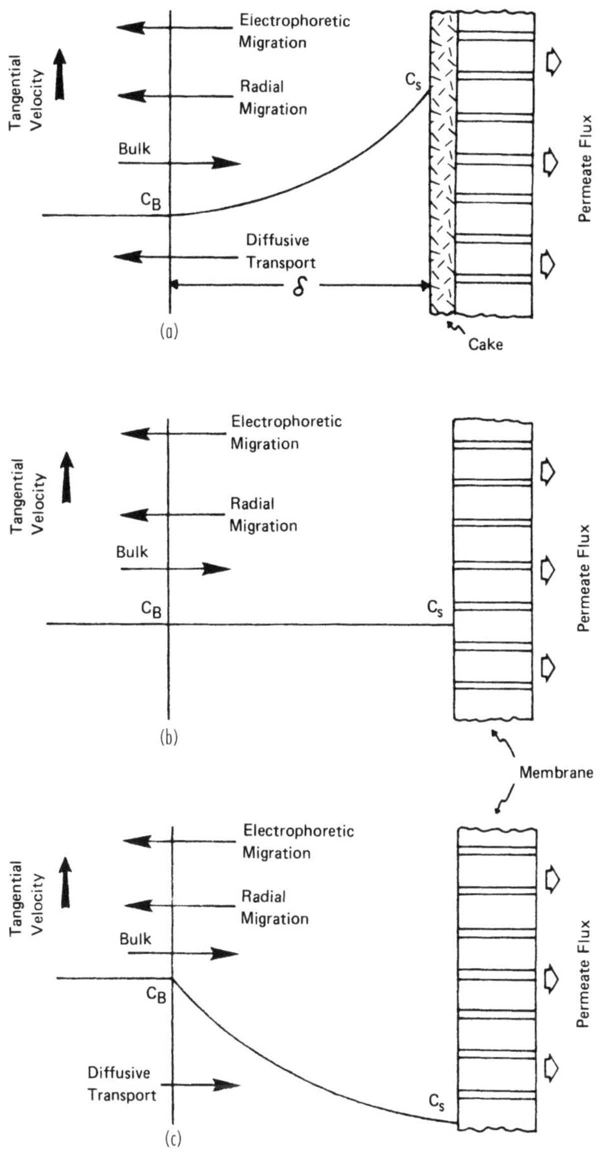

FIG. 20-27 Regimes of operation of cross-flow–electrofiltration: (*a*) voltage less than critical, (*b*) voltage equal to the critical voltage, (*c*) voltage greater than critical.

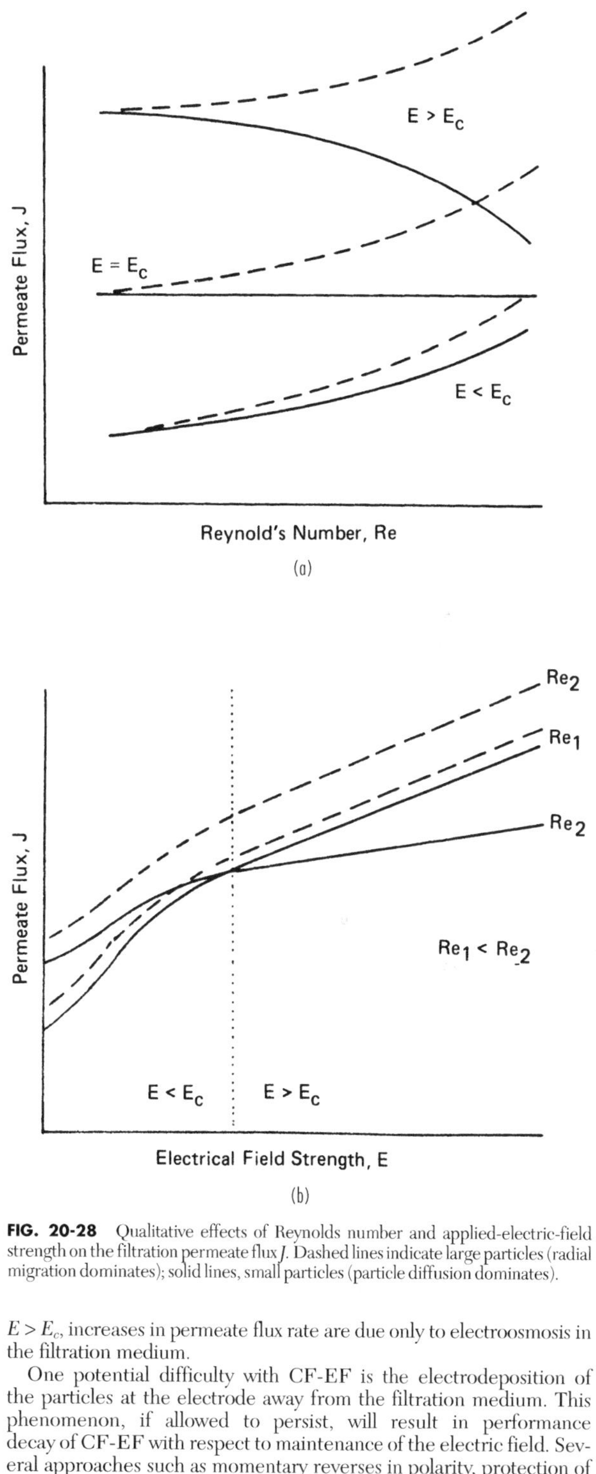

FIG. 20-28 Qualitative effects of Reynolds number and applied-electric-field strength on the filtration permeate flux *J*. Dashed lines indicate large particles (radial migration dominates); solid lines, small particles (particle diffusion dominates).

are observed. In this case, the electrophoretic-migration velocity away from the filter medium is greater than the velocity caused by bulk flow toward the filtration medium. Particles concentrate away from the filter medium. This implies that particle concentration is lowest next to the filter medium (in actuality, a clear boundary layer has been observed). The influence of fluid shear still improves the transfer of particles down the concentration gradient, but in this case it is toward the filtration medium. When the particles are small and diffusive transport dominates radial migration, increasing the circulation velocity will decrease the permeate flux rate in this regime. When the particles are large and radial migration dominates, the increase in circulation velocity will still improve the filtration rate. These effects are illustrated qualitatively in Fig. 20-28*a*. The solid lines represent systems in which the particle diffusive effect dominates the radial-migration effect, while the dashed lines represent the inverse. Figure 20-28*b* illustrates the increase in filtration rate with increasing electric field strength. For field strengths $E > E_c$, increases in permeate flux rate are due only to electroosmosis in the filtration medium.

One potential difficulty with CF-EF is the electrodeposition of the particles at the electrode away from the filtration medium. This phenomenon, if allowed to persist, will result in performance decay of CF-EF with respect to maintenance of the electric field. Several approaches such as momentary reverses in polarity, protection of the electrode with a porous membrane or filter medium, and/or utilization of a high fluid shear rate can minimize electrodeposition.

Dielectrophoresis

General References: Gascoyne and Vykoukal, *Electrophoresis,* **23,** 1973–1983 (2002). Pohl, in Moore (ed.), *Electrostatics and Its Applications,*

Wiley, New York, 1973, chap. 14 and chap. 15 (with Crane). Pohl, in Catsimpoolas (ed.), *Methods of Cell Separation,* vol. I, Plenum Press, New York, 1977, chap. 3. Pohl, *Dielectrophoresis: The Behavior of Matter in Nonuniform Electric Fields,* Cambridge, New York, 1978.

Introduction Dielectrophoresis (DEP) is defined as the motion of neutral, polarizable matter produced by a nonuniform electric (ac or dc) field. DEP should be distinguished from electrophoresis, which is the motion of charged particles in a uniform electric field (Fig. 20-29).

The DEP of numerous particle types has been studied, and many applications have been developed. Particles studied have included aerosols, glass, minerals, polymer molecules, living cells, and cell organelles. Applications developed include filtration, orientation, sorting or separation, characterization, and levitation and materials handling. Effects of DEP are easily exhibited, especially by large particles, and can be applied in many useful and desirable ways. DEP effects can, however, be observed on particles ranging in size even down to the molecular level in special cases. Since thermal effects tend to disrupt DEP with molecular-sized particles, they can be controlled only under special conditions such as in molecular beams.

Principle The principle of particle and cell separation, control, or characterization by the action of DEP lies in the fact that a net force can arise upon even neutral particles situated in a nonuniform electric field. The force can be thought of as rising from the imaginary two-step process of (1) induction or alignment of an electric dipole in a particle placed in an electric field followed by (2) unequal forces on the ends of that dipole. This arises from the fact that the force of an electric field upon a charge is equal to the amount of the charge and to the *local* field strength at that charge. Since the two (equal) charges of the (induced or oriented) dipole of the particle lie in unequal field strengths of the diverging field, a net force arises. If the particle is suspended in a fluid, then the polarizability of that medium enters, too.

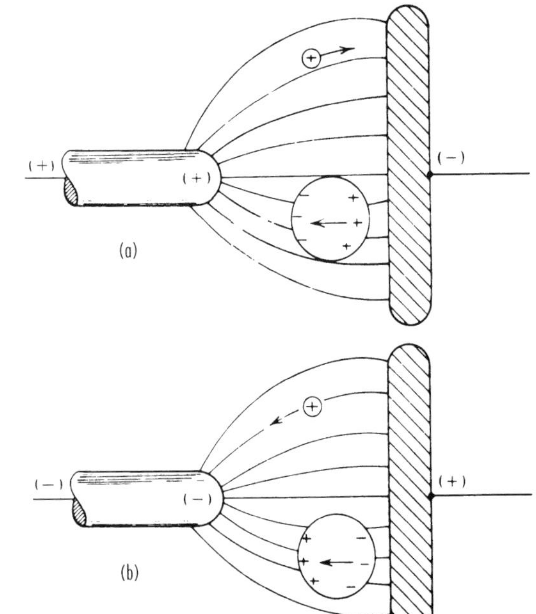

FIG. 20-29 Comparison of behaviors of neutral-charged bodies in an alternating nonuniform electric field. (*a*) Positively charged body moves toward negative electrode. Neutral body is polarized, then is attracted toward point where field is strongest. Since the two charge regions on the neutral body are equal in amount of charge but the force is proportional to the local field, a net force toward the region of more intense field results. (*b*) Positively charged body moves toward the negative electrode. Again, the neutral body is polarized, but it does not reverse direction although the field is reversed. It still moves toward the region of highest field intensity.

If, for example, the particle is more polarizable than the fluid, then the net force is such as to impel the particle to regions of greater field strength. Note that this statement implies that the effect is independent of the absolute sign of the field direction. This is found to be the case. Even rapidly alternating (ac) fields can be used to provide *unidirectional* motion of the suspended particles.

Formal Theory A small neutral particle at equilibrium in a static electric field experiences a net force due to DEP that can be written as $\mathbf{F} = (\mathbf{p} \cdot \nabla)\mathbf{E}$, where $\mathbf{p}$ is the dipole moment vector and $\mathbf{E}$ is the external electric field. If the particle is a simple dielectric and is isotropically, linearly, and homogeneously polarizable, then the dipole moment can be written as $\mathbf{p} = \alpha v\mathbf{E}$, where α is the (scalar) polarizability, v is the volume of the particle, and $\mathbf{E}$ is the external field. The force can then be written as:

$$\mathbf{F} = \alpha v(\mathbf{E} \cdot \nabla)\mathbf{E} = \tfrac{1}{2}\alpha v \nabla|\mathbf{E}|^2 \qquad (20\text{-}32)$$

This force equation can now be used to find the force in model systems such as that of an ideal dielectric sphere (relative dielectric constant K_2) in an ideal perfectly insulating dielectric fluid (relative dielectric constant K_1). The force can now be written as

$$\mathbf{F} = 2\pi a^3 \varepsilon_0 K_1 \left(\frac{K_2 - K_1}{K_2 + 2K_1}\right) \nabla|\mathbf{E}|^2 \qquad (20\text{-}33)$$

(ideal dielectric sphere in ideal fluid).

Heuristic Explanation As we can see from Fig. 20-30, the DEP response of *real* (as opposed to perfect insulator) particles with frequency can be rather complicated. We use a simple illustration to account for such a response. The force is proportional to the difference between the dielectric permittivities of the particle and the surrounding medium. Since a part of the polarization in real systems is thermally activated, there is a delayed response which shows as a phase lag between **D,** the dielectric displacement, and **E,** the electric-field intensity. To take this into account we may replace the simple (absolute) dielectric constant ε by the complex (absolute) dielectric constant $\hat{\varepsilon} = \varepsilon' - i\varepsilon'' = \varepsilon' - i\sigma/w$, where ω is the angular frequency of the applied field. For treating spherical objects, for example, the replacement

$$F \propto \frac{\varepsilon_1(\varepsilon_2 - \varepsilon_1)}{\varepsilon_2 + 2\varepsilon_1} \rightarrow \mathrm{Re}\left\{\frac{\hat{\varepsilon}_1^{\circ}(\hat{\varepsilon}_2 - \hat{\varepsilon}_1)}{\hat{\varepsilon}_2 + 2\hat{\varepsilon}_1}\right\} \qquad (20\text{-}34)$$

can be made, where $\hat{\varepsilon}^{\circ}$ is the complex conjugate of $\hat{\varepsilon}$.

With this force expression for real dielectrics, we can now explain the complicated DEP response with the help of Fig. 20-30.

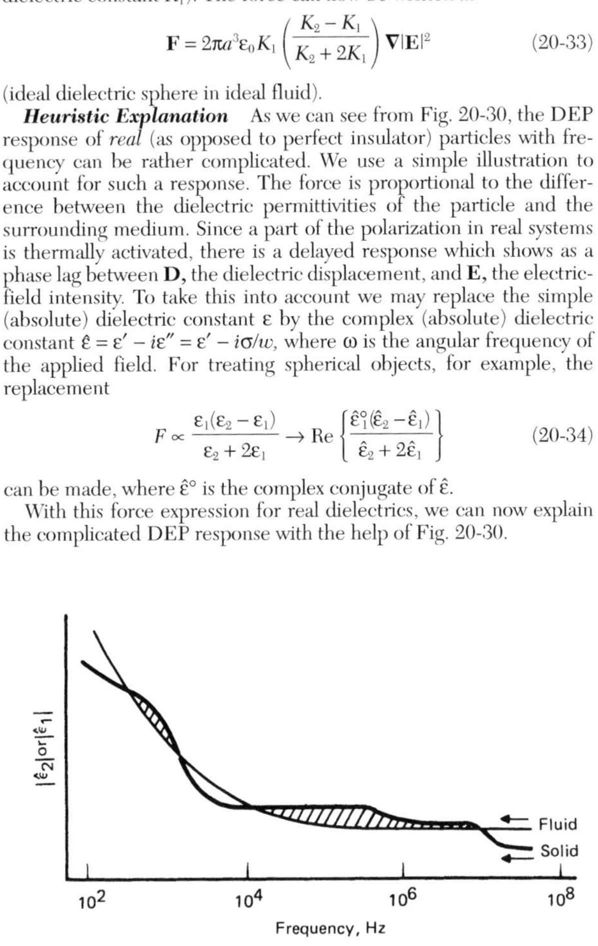

FIG. 20-30 A heuristic explanation of the dielectrophoretic-collection-rate (DCR)-frequency spectrum. The curves for the absolute values of the complex permittivities of the fluid medium and of the suspended particles are shown lying nearly, but not entirely, coincident over the frequency range of the applied electric field. When the permittivity (dielectric constant) of the particles exceeds that of the suspending medium, the collection, or "positive dielectrophoresis," occurs. In the frequency ranges in which the permittivity of the particles is less than that of the suspending medium no collection at the regions of higher field intensity occurs. Instead there is "negative dielectrophoresis," i.e., movement of the particles into regions of lower field intensity.

A particle, such as a living cell, can be imagined as having a number of different frequency-dependent polarization mechanisms contributing to the total effective polarization of the particle $|\hat{\varepsilon}_2|$. The heavy curve in Fig. 20-30 shows that the various mechanisms in the particle drop out stepwise as the frequency increases. The light curve in Fig. 20-30 shows the polarization for a simple homogeneous liquid that forms the surrounding medium. This curve is a smooth function which becomes constant at high frequency. As the curves cross each other (and hence $|\varepsilon_2| = |\varepsilon_1|$), various responses occur. The particle can thus be attracted to the strongest field region, be repelled from that region, or experience no force depending on the frequency.

Limitations It is desirable to have an estimate for the smallest particle size that can be effectively influenced by DEP. To do this, we consider the force on a particle due to DEP and also due to the osmotic pressure. This latter diffusional force will randomize the particles and tend to destroy the control by DEP. Figure 20-31 shows a plot of these two forces, calculated for practical and representative conditions, as a function of particle radius. As we can see, the smallest particles that can be effectively handled by DEP appear to be in range of 0.01 to 0.1 μm (100 to 1000 Å).

Another limitation to be considered is the volume that the DEP force can affect. This factor can be controlled by the design of electrodes. As an example, consider electrodes of cylindrical geometry. A practical example of this would be a cylinder with a wire running down the middle to provide the two electrodes. The field in such a system is proportional to $1/r$. The DEP force is then $\mathbf{F}_{DEP} \propto \nabla|E^2| \propto 1/r^3$, so that any differences in particle polarization might well be masked merely by positional differences in the force. At the outer cylinder the DEP force may even be too small to affect the particles appreciably. The most desirable electrode shape is one in which the force is independent of position within the nonuniform field. This "isomotive" electrode system is shown in Fig. 20-32.

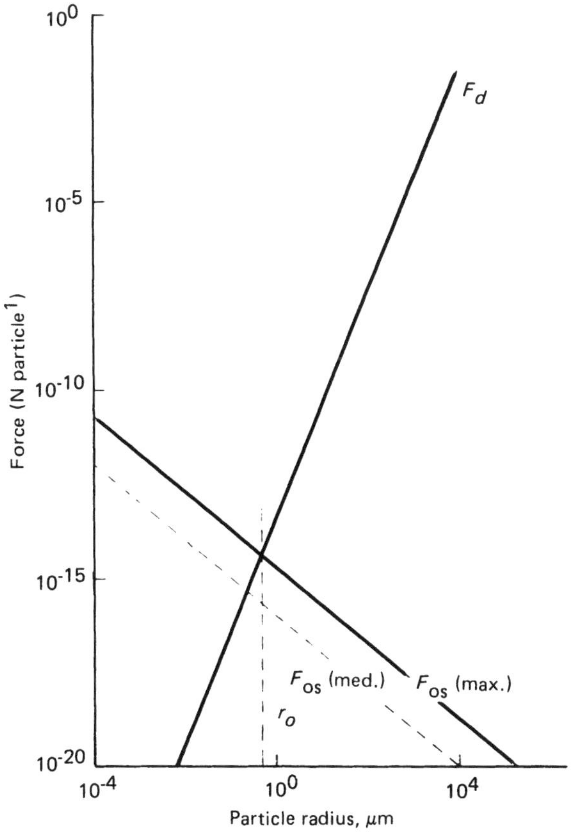

FIG. 20-31 Comparison of the dielectrophoretic (F_d) and osmotic (F_{os}) forces as functions of the particle size.

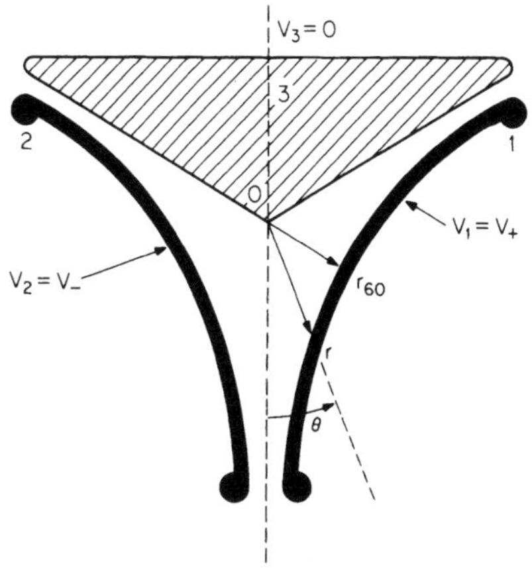

FIG. 20-32 A practical isomotive field geometry, showing r_{60}, the critical radius characterizing the isomotive electrodes. Electrode 3 is at ground potential, while electrodes 1 and 2 are at $V_1 = V_+$ and $V_2 = V_- = -V_+$ respectively. The inner faces of electrodes 1 and 2 follow $r = r_0\,[\sin(3\theta/2)]^{-2/3}$, while electrode 3 forms an angle of 120° about the midline.

Applications of Dielectrophoresis Over the past 20 years the use of DEP has grown rapidly to a point at which it is in use for biological [Hughes, *Electrophoresis*, **23,** 2569–2582 (2002)], colloidal, and mineral materials studies and handling. The effects of nonuniform electric fields are used for handling particulate matter far more often than is usually recognized. This includes the removal of particulate matter by "electrofiltration," the sorting of mixtures, or its converse, the act of mixing, as well as the coalescence of suspensions. In addition to these effects involving the translational motions of particles, some systems apply the orientational or torsional forces available in nonuniform fields. One well-known example of the latter is the placing of "tip-up" grit on emery papers commercially. Xerography and many other imaging processes are examples of multibillion-dollar industries which depend upon DEP for their success.

A clear distinction between **electrophoresis** (field action on an object carrying excess free charges) and **dielectrophoresis** (field gradient action on neutral objects) must be borne in mind at all times.

A dielectrofilter [Lin and Benguigui, *Sep. Purif. Methods*, **10**(1), 53 (1981); Sisson et al., *Sep. Sci. Technol.*, **30**(7–9), 1421 (1995)] is a device which uses the action of an electric field to aid the filtration and removal of particulates from fluid media. A dielectrofilter can have a very obvious advantage over a mechanical filter in that it can remove particles which are much *smaller* than the flow channels in the filter. In contrast, the ideal mechanical filter must have all its passages smaller than the particles to be removed. The resultant flow resistance can be use-restrictive and energy-consuming unless a phenomenon such as dielectrofiltration is used.

Dielectrofiltration can (and often does) employ both electrophoresis and dielectrophoresis in its application. The precise physical process which dominates depends on a number of physical parameters of the system. Factors such as field intensity and frequency and the electrical conductivity and dielectric constants of the materials present determine this. Although these factors need constant attention for optimum operation of the dielectrofilter, this additional complication is often more than compensated for by the advantages of dielectrofiltration such as greater throughput and lesser sensitivity to viscosity problems, etc. To operate the dielectrofilter in the dominantly electrophoretic mode requires that excess free charges of one sign or the other reside on the particulate matter. The necessary charges can be those naturally present, as upon a charged sol; or they may need to be artificially implanted such as by passing the particles through a corona discharge.

Dielectrofiltration by the corona-charging, electrophoresis-dominated Cottrell technique is now widely used.

To operate the dielectrofilter (dominantly dielectrophoretic mode), on the other hand, one must avoid the presence of free charge on the particles. If the particles can become charged during the operation, a cycle of alternate charging and discharging in which the particles dash to and from the electrodes can occur. This is most likely to occur if static or very low frequency fields are used. For this reason, corona and like effects may be troublesome and need often to be minimized. To be sure, the DEP force is proportional to the field applied [actually to $\nabla(E)^2$], but fields which are too intense can produce such troublesome charge injection. A compromise for optimal operation is necessary between having $\nabla(E)^2$ so low that DEP forces are insufficient for dependable operation, on the one hand, and having **E** so high that troublesome discharges (e.g., coronalike) interfere with dependable operation of the dielectrofilter. In insulative media such as air or hydrocarbon liquids, for example, one might prefer to operate with fields in the range of, say, 10 to 10,000 V/cm. In more conductive media such as water, acetone, or alcohol, for example, one would usually prefer rather lower fields in the range of 0.01 to 100 V/cm. The higher field ranges cited might become unsuitable if conductive sharp asperities are present.

Another factor of importance in dielectrofiltration is the need to have the DEP effect firmly operative upon *all* portions of the fluid passing through. Oversight of this factor is a most common cause of incomplete dielectrofiltration. Good dielectrofilter design will emphasize this crucial point. To put this numerically, let us consider the essential field factor for DEP force, namely $\nabla(\mathbf{E}_0)^2$. Near sharp points, e.g., **E,** the electric field varies with the radial distance r as $E \propto r^{-2}$; hence our DEP force factor will vary as $\nabla(E)^2 \propto r^{-5}$. In the neighborhood of sharp "line" sources such as at the edge of electrode plates, $E \propto r^{-1}$, hence, $\nabla(E)^2 \propto r^{-3}$. If, for instance, the distance is varied by a factor of 4 from the effective field source in these cases, the DEP force can be expected to weaken by a factor of 1024 or 64 respectively for the point source and the line source. The matter is even more keenly at issue when field-warping dielectrics (defined later) are used to effect maximal filtration. In this case the field-warping material is made to produce dipole fields as induced by the applied electric field. If we ask how the crucial factor, $\nabla(E)^2$, varies with distance away from such a dipole, we find that since the field E_d about a dipole varies approximately as r^{-3}, then $\nabla(E)^2$ can be expected to vary as r^{-7}. It then becomes critically important that the particles to be removed from the passing fluid do, indeed, pass very close to the surface of the field-warping material, or it will not be effectively handled. Clearly, it would be difficult to maintain successfully uniform dielectrofiltration treatment of fluid passing through such wildly variant regions. The problems can be minimized by ensuring that all the elements of the passing fluid go closely by such field sources in the dielectrofilter. In practice this is done by constructing the dielectrofilter from an assembly of highly comminuted electrodes or else by a set of relatively simple and widely spaced metallic electrodes between which is set an assembly of more or less finely divided solid dielectric material having a complex permittivity different from that of the fluid to be treated. The solid dielectric (fibers, spheres, chunks) serves to produce field nonuniformities or **field warpings** to which the particles to be filtered are to be attracted. In treating fluids of low dielectric constant such as air or hydrocarbon fluids, one sees field-warping materials such as sintered ceramic balls, glass-wool matting, open-mesh polyurethane foam, alumina, chunks, or $BaTiO_3$ particles.

An example of a practical dielectrofilter which uses both of the features described, namely, sharp electrodes *and* dielectric field-warping filler materials, is that described in Fig. 20-33 [H. J. Hall and R. F. Brown, *Lubric. Eng.*, **22,** 488 (1966)]. It is intended for use with hydraulic fluids, fuel oils, lubricating oils, transformer oils, lubricants, and various refinery streams. Performance data are cited in Fig. 20-34. It must be remarked that in the opinion of Hall and Brown the action of the dielectrofilter was "electrostatic" and due to free charge on the particles dispersed in the liquids. It is the present authors' opinion, however, that both electrophoresis and dielectrophoresis are operative here but that the dominant mechanism is that of DEP, in which neutral particles are polarized and attracted to the regions of highest field intensity.

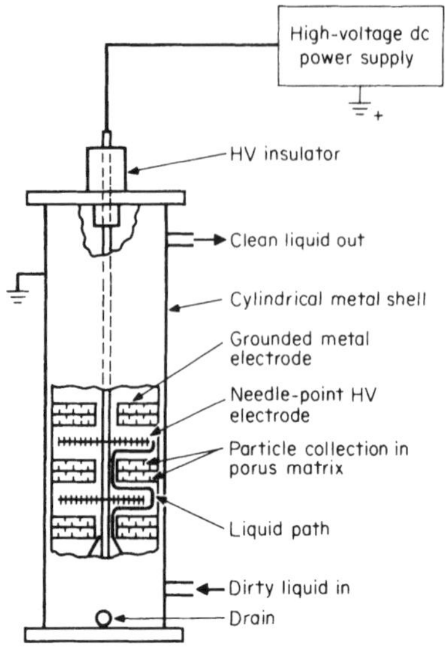

FIG. 20-33 Diagram illustrating the function of an electrostatic liquid cleaner.

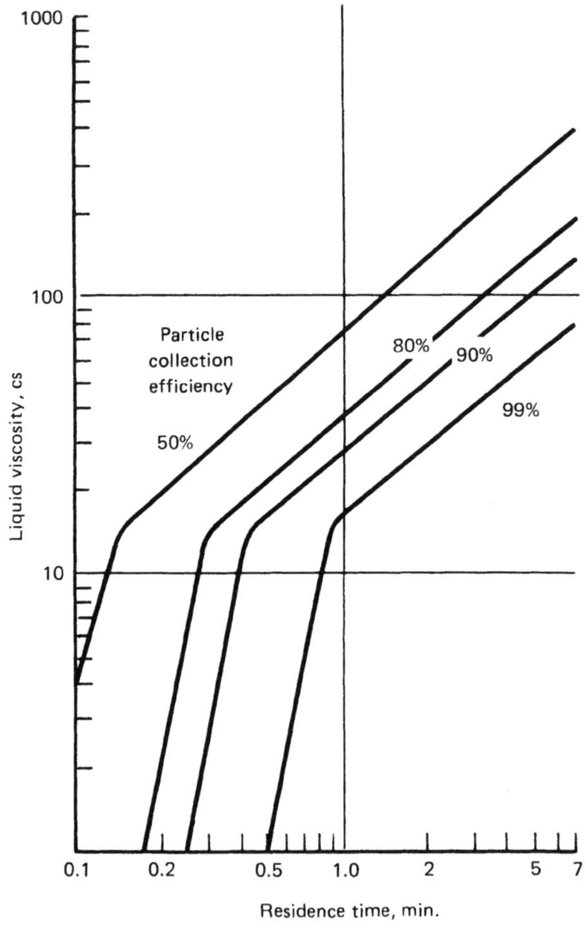

FIG. 20-34 Performance data for a typical high-efficiency electrostatic liquid cleaner.

A second commercial example of dielectric filtration is the Gulftronic® separator [G. R. Fritsche, *Oil & Gas J.*, **75**, 73 (1977)] which was commercialized in the late 1970s by Gulf Science and Technology Company. Instead of using needle-point electrodes as shown in Fig. 20-33, the Gulftronic® separator relied on the use of a bed of glass beads to produce the field nonuniformities required for dielectric filtration. Either ac or dc electric fields could be used in this separator. The Gulftronic® separator has been used primarily to remove catalyst fines from FCC decant oils and has been reported to exhibit removal efficiencies in excess of 80 percent for this fine-particle separation problem.

Another example of the commercial use of DEP is in polymer clarification [A. N. Wennerberg, U.S. Patent 2,914,453, 1959; assignor to Standard Oil Co. (Indiana)]. Here, either ac or dc potentials were used while passing suspensions to be clarified through regions with an area-to-electrode-area ratio of 10:1 or 100:1 and with fields in the order of 10 kV/cm. Field warping by the presence of various solid dielectrics was observed to enhance filtration considerably, as expected for DEP. The filtration of molten or dissolved polymers to free them of objectionable quantities of catalyst residues, for example, was more effective if a solid dielectric material such as Attapulgus clay, silica gel, fuller's earth, alumina, or bauxite was present in the region between the electrodes. The effectiveness of percolation through such absorptive solids for removing color bodies is remarkably enhanced by the presence of an applied field. A given amount of clay is reported to remove from 4 to 10 times as much color as would be removed in the absence of DEP. Similar results are reported by Lin et al. [Lin, Yaniv, and Zimmels, *Proc. XIIIth Int. Miner. Process. Congr.*, Wrocław, Poland, 83–105 (1979)].

The instances cited were examples of the use of DEP to filter liquids. We now turn to the use of DEP to aid in dielectrofiltration of gases. Fielding et al. observe that the effectiveness of high-quality fiberglass air filters is dramatically improved by a factor of 10 or more by incorporating DEP in the operation. Extremely little current or power is required, and no detectable amounts of ozone or corona need result. The DEP force, once it has gathered the particles, continues to act on the particles already sitting on the filter medium, thereby improving adhesion and minimizing blowoff.

The degree by which the DEP increases the effectiveness of gas filtration, or the dielectrophoretic augmentation factor (DAF), is definable. It is the ratio of the volumes of aerosol-laden gas which can be cleaned effectively by the filter with and without the voltage applied. For example, the application of 11 kV/cm gave a DAF of 30 for 1.0-μm-diameter dioctyl phthalate particles in air, implying that the penetration of the glass filter is reduced thirtyfold by the application of a field of 1100 kV/m. Similar results were obtained by using "standard" fly ash supplied by the Air Pollution Control Office of the U.S. Environmental Protection Agency. The data obtained for several aerosols tested are shown in Table 20-14 and in Fig. 20-35. The relation DAF = kV^2/v is observed to hold approximately for each aerosol. Here, the DEP augmentation factor DAF is observed to depend upon a constant K, a characteristic of the material, upon the square of the applied voltage, and upon the inverse of the volume flow rate v through the filter.

It is worth noting that in the case of the air filter described DEP serves as an augmenting rather than as an exclusive mechanism for the removal of particulate material. It is a unique feature of the dielectrophoretic gas filter that the DEP force is maximal when the particulates are at or on the fiber surface. This causes the deposits to be strongly retained by this particular filtration mechanism. It thus contrasts importantly with other types of gas filter in which the filtration mechanism no longer acts after the capture of the particle. In particular, in the case of the older electrostatic mechanisms involving only coulombic attraction, a simple charge alternation on the particle, such as caused by normal conduction, often evokes disruption of the filter operation because of particle repulsion from the contacting electrode. On the other hand, ordinary mechanical filtration depends upon the action of adventitious particle trapping or upon van der Waals forces, etc., to hold the particles. The high efficiency possible with electrofilters suggests their wider use.

TABLE 20-14 Dielectrophoretic Augmentation of Filtration of a Liquid Aerosol*

Air speed, cm/s	DAF at			
	2 kV	3.5 kV	5 kV	7 kV
0.3-μm-diameter dioctyl phthalate aerosol				
3	8	19	95	330
6	3	13	39	120
9	3	11	28	100
15	2	6	13	42
20	2	5	9	27
28	2	4	6	14
39	2	3	4	9
50	1	2	3	6
1.0-μm-diameter dioctyl phthalate aerosol				
3	30	110	300	1100
6	6	3	95	360
9	4	18	50	170
15	3	10	20	50
20	2	6	13	35
28	2	4	8	18
39	2	3	5	11
50	1	2	3	7
Fly-ash aerosol				
6	10	30	80	
10	8	30	80	
14	5	20	40	
20	4	10	30	70
35	3	7	10	20
45	1	2	6	
53	1	2	7	10

*Experimentally measured dielectrophoretic augmentation factor DAF as a function of air speed and applied voltage for a glass-fiber filter (HP-100, Farr Co.). Cf. Fielding, Thompson, Bogardus, and Clark, *Dielectrophoretic Filtration of Solid and Liquid Aerosol Particulates*, Prepr. 75-32.2, 68th ann. meet., Air Pollut. Control Assoc., Boston, June 1975.

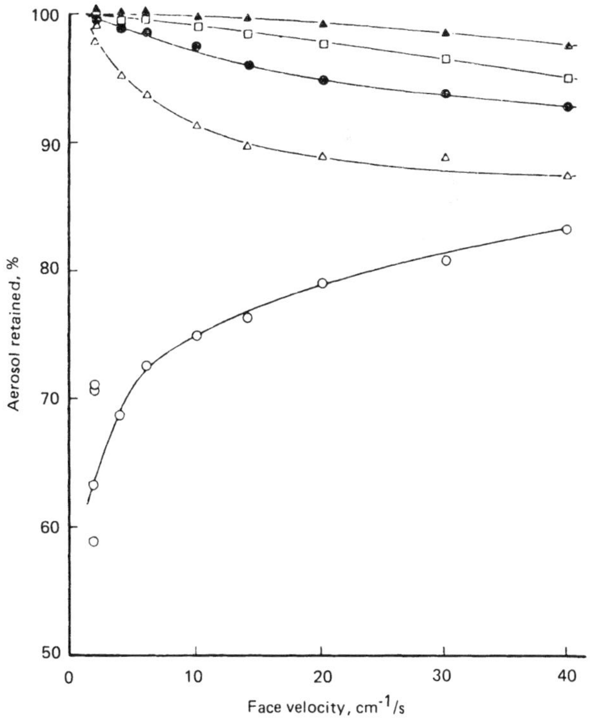

FIG. 20-35 Efficiency of an electrofilter as a function of gas flow rate at 5 different voltages. Experimental materials: 1-μm aerosol of dioctyl phthalate; glass-fiber filter. Symbols: ○, no voltage applied; Δ, 2 kV; ●, 3.5 kV; □, 5 kV; ▲, 7 kV. (*After Fielting et al.,* Dielectrophoretic Filtration of Solid and Liquid Aerosol Particulates, *Prepr. 75-32.2, 68th ann. meet., Air Pollut. Control Assoc., Boston, June 1975.*)

SURFACE-BASED SOLID-LIQUID SEPARATIONS INVOLVING A SECOND LIQUID PHASE

GENERAL REFERENCES: Fuerstenau, "Fine Particle Flotation," in Somasundaran (ed.), *Fine Particles Processing*, vol. 1, American Institute of Mining, Metallurgical, and Petroleum Engineers, New York, 1980. Henry, Prudich, and Lau, *Colloids Surf.*, **1,** 335 (1980). Henry, Prudich, and Vaidyanathan, *Sep. Purif. Methods*, **8**(2), 31 (1979). Jacques, Hovarongkura, and Henry, *Am. Inst. Chem. Eng. J.*, **25**(1), 160 (1979). Stratton-Crawley, "Oil Flotation: Two Liquid Flotation Techniques," in Somasundaran and Arbiter (eds.), *Beneficiation of Mineral Fines*, American Institute of Mining, Metallurgical, and Petroleum Engineers, New York, 1979.

Process Concept Three potential surface-based regimes of separation exist when a second, immiscible liquid phase is added to another, solids-containing liquid in order to effect the removal of solids. These regimes (Fig. 20-36) are:

1. Distribution of the solids into the bulk second liquid phase
2. Collection of the solids at the liquid-liquid interface
3. Bridging or clumping of the solids by the added fluid in order to form an agglomerate followed by settling or filtration

These separation techniques should find particular application in systems containing fine particles. The surface chemical differences involved among these separation regimes are only a matter of degree; i.e., all three regimes require the wetting of the solid by the second liquid phase. The addition of a surface-active agent is sometimes needed in order to achieve the required solids wettability. In spite of this similarity, applied processing (equipment configuration, operating conditions, etc.) can vary widely. Collection at the interface would normally be treated as a flotation process (see also "Adsorptive-Bubble Separation Methods" in Sec. 20), distribution to the bulk liquid as a liquid-liquid extraction analog, and particle bridging as a settling (sedimentation) or filtration process.

Even though surface-property-based liquid-solid-liquid separation techniques have yet to be widely used in significant industrial applications, several studies which demonstrate their effectiveness have appeared in literature.

Albertsson (*Partition of Cell Particles and Macromolecules*, 3d ed., Wiley, New York, 1986) has extensively used particle distribution to fractionate mixtures of biological products. (See also Sec. 15.) In order to demonstrate the versatility of particle distribution, he has cited the example shown in Table 20-15. The feed mixture consisted of polystyrene particles, red blood cells, starch, and cellulose. Liquid-liquid particle distribution has also been studied by using mineral-matter particles (average diameter = 5.5 μm) extracted from a coal liquid as the solid in a xylene-water system [Prudich and Henry, *Am. Inst. Chem. Eng. J.*, **24**(5), 788 (1978)]. By using surface-active agents in order to enhance the water wettability of the solid particles, recoveries of better than 95 percent of the particles to the water phase were observed. All particles remained in the xylene when no surfactant was added.

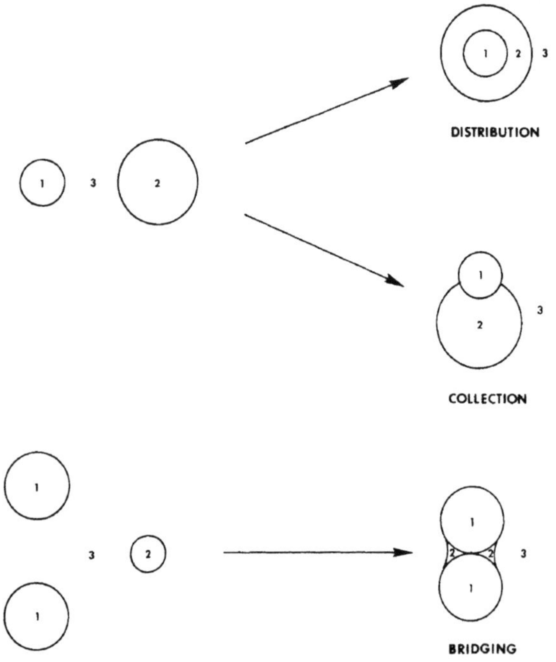

FIG. 20-36 Regimes of separation in a liquid-solid-liquid system. Phase 1 = particle; phase 2 = liquid (dispersed); phase 3 = liquid (continuous).

TABLE 20-15 Separations of Particles between Two Phases

System	Top phase	Bottom phase
Polyethylene glycol / salt	Polystyrene	All others
PEG / Dextran; 20,000 MW	Algae	All others
PEG / Dextran; 200,000 MW	Red cells	All others
Methyl cellulose / Dextran	Cellulose particles	Starch

Particle collection at a liquid-liquid interface is a particularly favorable separation process when applied to fine-particle systems. Advantages of this type of processing include:

- Decreased liquid-liquid interfacial tension (when compared with a gas-liquid system) results in higher liquid-liquid interfacial areas, which favor solid-particle droplet collisions.
- Liquid-solid interactions due to long-range intermolecular forces are much larger than are gas-solid interactions. This means that it is easier to collect fine particles at a liquid-liquid interface than at a gas-liquid interface.
- The increased momentum of liquid droplets (when compared with gas) should favor solid-particle collection.

Fuerstenau [Lai and Fuerstenau, *Trans. Am. Inst. Min. Metall. Pet. Eng.*, **241,** 549 (1968); Raghavan and Fuerstenau, *Am. Inst. Chem. Eng. Symp. Ser.*, **71**(150), 59 (1975)] has studied this process with respect to the removal of alumina particles (0.1 μm) and hematite particles (0.2 μm) from an aqueous solution by using isooctane. The use of isooctane as the collecting phase for the hematite particles resulted in an increase in particle recovery of about 50 percent over that measured when air was used as the collecting phase under the same conditions. The effect of the wettability of the solid particles (as measured by the three-phase contact angle) on the recovery of hematite in the water-isooctane system is shown in Fig. 20-37. This behavior is typical of particle collection. Particle collection at an oil-water interface has also been studied with respect to particle removal from a coal liquid. Particle removals averaging about 80 percent have been observed when water is used as the collecting phase (Lau, master's thesis, West Virginia University, 1979). Surfactant addition was necessary in order to control the wettability of the solids.

Particle bridging has been chiefly investigated with respect to spherical agglomeration. Spherical agglomeration involves the collecting or transferring of the fine particles from suspension in a liquid phase into spherical aggregates held together by a second liquid phase. The aggregates are then removed from the slurry by filtration or settling. Like the other liquid-solid-liquid separation techniques, the solid must be wet by the second liquid phase. The spherical agglomeration process has resulted in the development of a pilot unit called the Shell Pelletizing Separator [Zuiderweg and Van Lookeren Campagne, *Chem. Eng.* (*London*), **220,** CE223 (1968)].

The ability to determine in advance which of the separation regimes is most advantageous for a given liquid-solid-liquid system would be desirable. No set of criteria with which to make this determination presently exists. Work has been done with respect to the

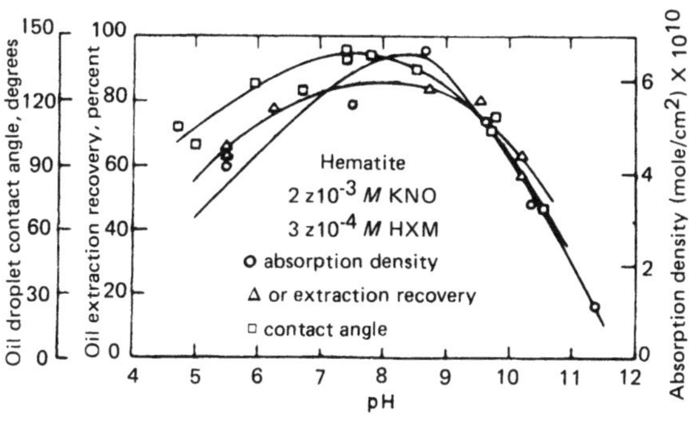

FIG. 20-37 The variation of adsorption density, oil-droplet contact angle, and oil-extraction recovery of hematite as a function of pH. To convert gram-moles per square centimeter to pound-moles per square foot, multiply by 2.048. [*From Raghavan and Fuerstenau,* Am. Inst. Chem. Eng. Symp. Ser., *71(150), 59 (1975).*]

identification of system parameters which make these processes technically feasible. The results of these studies can be used to guide the selection of the second liquid phase as well as to suggest approximate operating conditions (dispersed-liquid droplet size, degree and type of mixing, surface-active-chemical addition, etc.).

Theory Theoretical analyses of spherical particles suspended in a planar liquid-liquid interface have appeared in literature for some time, the most commonly presented forms being those of a free energy and/or force balance made in the absence of all external body forces. These analyses are generally used to define the boundary criteria for the shift between the collection and distribution regimes, the bridging regime not being considered. This type of analysis shows that for a spherical particle possessing a three-phase contact angle between 0 and 180°, as measured through the receiving or collecting phase, collection at the interface is favored over residence in either bulk phase. These equations are summarized, using a derivation of Young's equation, as

$$\frac{\gamma_{s2} - \gamma_{s1}}{\gamma_{12}} > 1 \text{ particles wet to phase 1} \qquad (20\text{-}35)$$

$$\frac{\gamma_{s2} - \gamma_{s1}}{\gamma_{12}} < -1 \text{ particles wet to phase 2} \qquad (20\text{-}36)$$

$$\left|\frac{\gamma_{s2} - \gamma_{s1}}{\gamma_{12}}\right| \leq 1 \text{ particle at interface} \qquad (20\text{-}37)$$

where γ_{ij} is the surface tension between phases i and j, N/m (dyn/cm); s indicates the solid phase; and subscripts 1 and 2 indicate the two liquid phases.

Several additional studies [Winitzer, *Sep. Sci.*, **8**(1), 45 (1973); ibid., **8**(6), 647 (1973); Maru, Wasan, and Kintner, *Chem. Eng. Sci.*, **26**, 1615 (1971); and Rapacchietta and Neumann, *J. Colloid Interface Sci.*, **59**(3), 555 (1977)] which include body forces such as gravitational acceleration and buoyancy have been made. A typical example of a force balance describing such a system (Fig. 20-38) is summarized in Eq. (20-38):

$$[(\gamma_{s1} - \gamma_{s2}) \cos \delta + \gamma_{12} \cos B]L = g[V_{\text{total}}\rho_s - V_1\rho_1 - V_2\rho_2] \qquad (20\text{-}38)$$

where V_1 is the volume of the particle in fluid phase 1, V_2 is the volume in fluid phase 2, L is the particle circumference at the interface between the two liquid phases, ρ_i is the density of phase i, and g is the gravitational constant. The left-hand side of the equation represents the surface forces acting on the solid particle, while the right-hand side includes the gravitational and buoyancy forces. This example illustrates the fact that body forces can have a significant effect on system behavior. The solid-particle size as well as the densities of the solid and both liquid phases are introduced as important system parameters.

A study has also been performed for particle distribution for cases in which the radii of curvature of the solid and the liquid-liquid interface

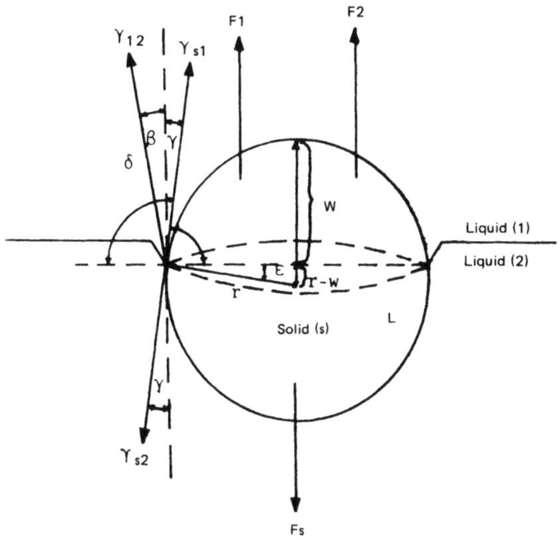

FIG. 20-38 Solid sphere suspended at the liquid-liquid interface. F_1 and F_2 are buoyancy forces; F_S is gravity. [*From Winitzer,* Sep. Sci., *8(1), 45 (1973).*]

are of the same order of magnitude [Jacques, Hovarongkura, and Henry, *Am. Inst. Chem. Eng. J.*, **25**(1), 160 (1979)]. Differences between the final and initial surface free energies are used to analyze this system. Body forces are neglected. Results (Fig. 20-39) demonstrate that n, the ratio of the particle radius to the liquid-liquid-interface radius, is an important system parameter. Distribution of the particle from one phase to the other is favored over continued residence in the original phase when the free-energy difference is negative. For a solid particle of a given size, these results show that as the second-phase droplet size decreases, the contact angle required in order to effect distribution decreases (the required wettability of the solid by the second phase increases). The case of particle collection at a curved liquid-liquid interface has also been studied in a similar manner [Smith and Van de Ven, *Colloids Surf.*, **2**, 387 (1981)]. This study shows that collection is preferred over distribution for any n in systems without external body forces when the contact angle lies between 0 and 180°.

While thermodynamic-stability studies can be valuable in evaluating the technical feasibility of a process, they are presently inadequate in determining which separation regime will dominate a particular liquid-solid-liquid system. These analyses ignore important processing phenomena such as the mechanism of encounter of the dispersed-phase liquid with the solid particles, the strength of particle attachment, and the mixing-energy input necessary to effect the separation. No models of good predictive value which take all these variables into account have yet been offered. Until the effects of these and other system variables can be adequately understood, quantified, and combined into such a predictive model, no a priori method of performance prediction will be possible.

ADSORPTIVE-BUBBLE SEPARATION METHODS

General References: Lemlich (ed.), *Adsorptive Bubble Separation Techniques*, Academic, New York, 1972. Carleson, "Adsorptive Bubble Separation Processes" in Scamehorn and Harwell (eds.), *Surfactant-Based Separation Processes*, Marcel Dekker, New York, 1989.

Principle The adsorptive-bubble separation methods, or adsubble methods for short [Lemlich, *Chem. Eng.* **73**(21), 7 (1966)], are based on the selective adsorption or attachment of material on the surfaces of gas bubbles passing through a solution or suspension. In most of the methods, the bubbles rise to form a foam or froth which carries the material off overhead. Thus the material (desirable or undesirable) is removed from the liquid, and not vice versa as in, say,

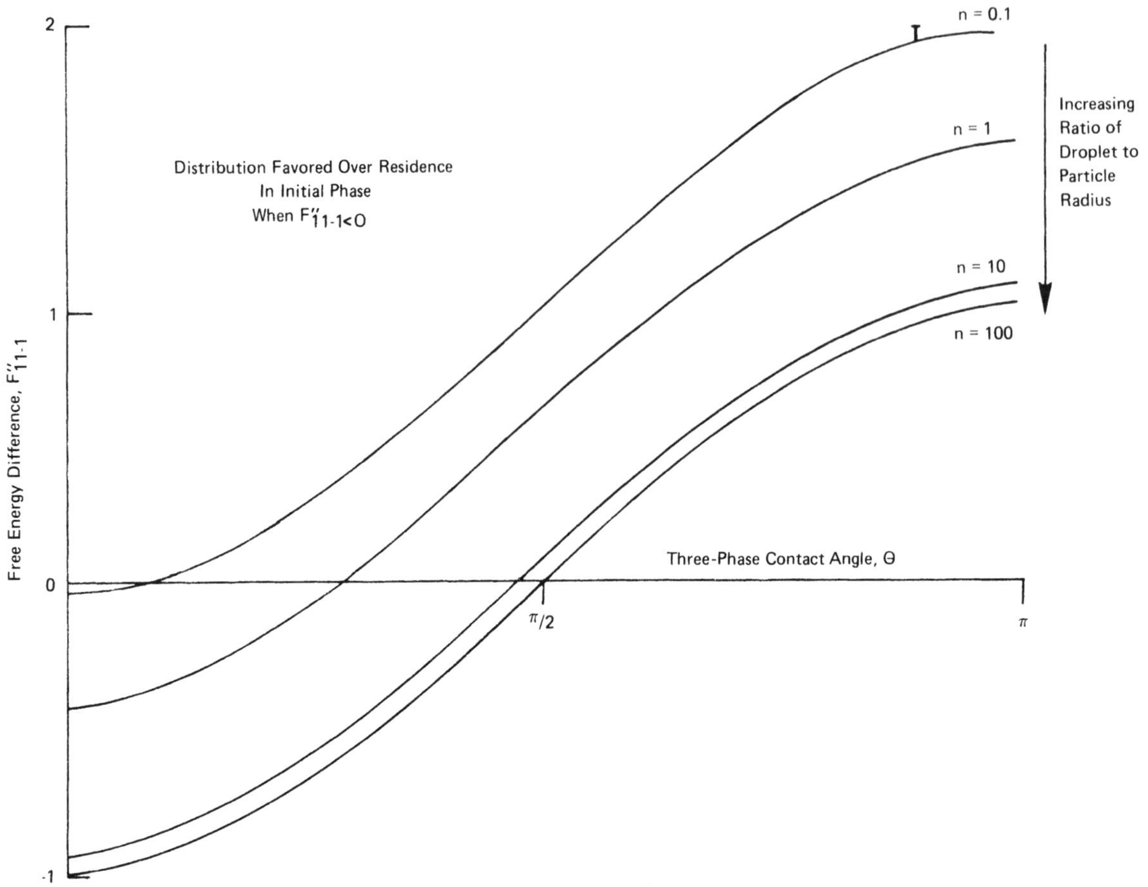

FIG. 20-39 Normalized free-energy difference between distributed (II) and nondistributed (I) states of the solid particles versus three-phase contact angle (collection at the interface is not considered). A negative free-energy difference implies that the distributed state is preferred over the nondistributed state. Note especially the significant effect of *n*, the ratio of the liquid droplet to solid-particle radius. [*From Jacques, Hovarongkura, and Henry,* Am. Inst. Chem. Eng. J., **25**(*1*), *160* (*1979*).]

filtration. Accordingly, the foaming methods appear to be particularly (although not exclusively) suited to the removal of small amounts of material from large volumes of liquid.

For any adsubble method, if the material to be removed (termed the **colligend**) is not itself surface-active, a suitable surfactant (termed the **collector**) may be added to unite with it and attach or adsorb it to the bubble surface so that it may be removed (Sebba, *Ion Flotation,* Elsevier, New York, 1962). The union between colligend and collector may be by chelation or other complex formation. Alternatively, a charged colligend may be removed through its attraction toward a collector of opposite charge.

Definitions and Classification Figure 20-40 outlines the most widely accepted classification of the various adsubble methods [Karger, Grieves, Lemlich, Rubin, and Sebba, *Sep. Sci.,* **2,** 401 (1967)]. It is based largely on actual usage of the terms by various workers, and so the definitions include some unavoidable inconsistencies and overlap.

Among the methods of foam separation, **foam fractionation** usually implies the removal of dissolved (or sometimes colloidal) material. The overflowing foam, after collapse, is called the *foamate.* The solid lines of Fig. 20-41 illustrate simple continuous foam fractionation. (Batch operation would be represented by omitting the feed and bottoms streams.)

On the other hand, **flotation** usually implies the removal of solid particulate material. Most important under the latter category is **ore flotation.**

Also under the category of flotation are to be found **macroflotation,** which is the removal of macroscopic particles; **microflotation** (also called **colloid flotation**), which is the removal of microscopic particles, particularly colloids or microorganisms [Dognon and Dumontet, *Comptes Rendus,* **135,** 884 (1941)]; **molecular flotation,** which is the removal of surface-inactive molecules through the use of a collector (surfactant) which yields an insoluble product; **ion flotation,** which is the removal of surface-inactive ions via a collector which yields an insoluble product, especially a removable scum [Sebba, *Nature,* **184,** 1062 (1959)]; **adsorbing colloid flotation,** which is the removal of dissolved material in piggyback fashion by adsorption on colloidal particles; and **precipitate flotation,** in which a precipitate is removed by a collector which is not the precipitating agent [Baarson and Ray, "Precipitate Flotation," in Wadsworth and Davis (eds.), *Unit Processes in Hydrometallurgy,* Gordon and Breach, New York, 1964, p. 656]. The last definition has been narrowed to precipitate flotation of the first kind, the second kind requiring no separate collector at all [Mahne and Pinfold, *J. Appl. Chem.,* **18,** 52 (1968)].

A separation can sometimes be obtained even in the absence of any foam (or any floated floc or other surrogate). In **bubble fractionation** this is achieved simply by lengthening the bubbled pool to form a vertical column [Dorman and Lemlich, *Nature,* **207,** 145 (1965)]. The ascending bubbles then deposit their adsorbed or attached material at the top of the pool as they exit. This results in a concentration gradient which can serve as a basis for separation. Bubble fractionation can operate either alone or as a booster section below a foam

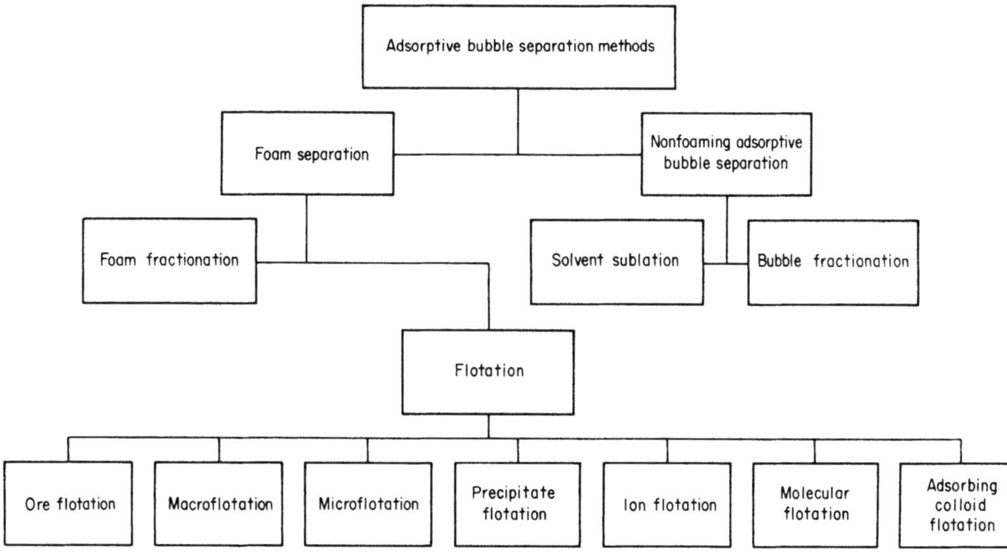

FIG. 20-40 Classification for the adsorptive-bubble separation methods.

fractionator, perhaps to raise the concentration up to the foaming threshold.

In **solvent sublation** an immiscible liquid is placed atop the main liquid to trap the material deposited by the bubbles as they exit (Sebba, *Ion Flotation,* Elsevier, New York, 1962). The upper liquid should dissolve or at least wet the material. With appropriate selectivity, the separation so achieved can sometimes be much greater than that with bubble fractionation alone.

The droplet analogs to the adsubble methods have been termed the **adsoplet methods** (from adsorptive droplet separation methods) [Lemlich, "Adsorptive Bubble Separation Methods," *Ind. Eng. Chem.,* **60**(10), 16 (1968)]. They are omitted from Fig. 20-40, since they involve adsorption or attachment at *liquid-liquid* interfaces. Among them are **emulsion fractionation** [Eldib, "Foam and Emulsion Fractionation," in Kobe and McKetta (eds.), *Advances in Petroleum Chemistry and Refining,* vol. 7, Interscience, New York, 1963, p. 66], which is the analog of foam fractionation; and **droplet fractionation** [Lemlich, loc. cit.; and Strain, *J. Phys. Chem.,* **57,** 638 (1953)], which is the analog of bubble fractionation. Similarly, the old beneficiation operation called bulk oil flotation (Gaudin, *Flotation,* 2d ed., McGraw-Hill, New York, 1957) is the analog of modern ore flotation. By and large, the adsoplet methods have not attracted the attention accorded to the adsubble methods.

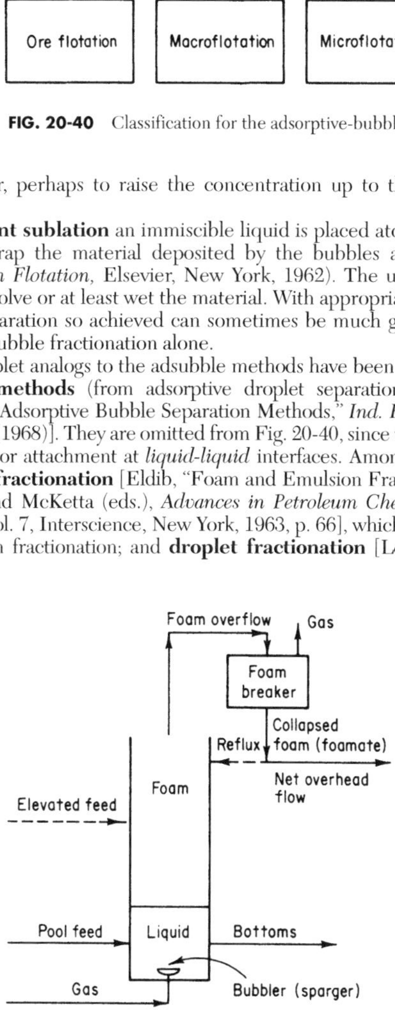

FIG. 20-41 Four alternative modes of continuous-flow operation with a foam-fractionation column: (1) The simple mode is illustrated by the solid lines. (2) Enriching operation employs the dashed reflux line. (3) In stripping operation, the elevated dashed feed line to the foam replaces the solid feed line to the pool. (4) For combined operation, reflux and elevated feed to the foam are both employed.

Of all the adsubble methods, foam fractionation is the one for which chemical engineering theory is the most advanced. Fortunately, some of this theory also applies to other adsubble methods.

Adsorption The separation achieved depends in part on the selectivity of adsorption at the bubble surface. At equilibrium, the adsorption of dissolved material follows the Gibbs equation (Gibbs, *Collected Works,* Longmans Green, New York, 1928).

$$d\gamma = -\mathbf{R}T\Sigma\Gamma_i d\ln a_i \qquad (20\text{-}39)$$

Γ_i is the surface excess (Davies and Rideal, *Interfacial Phenomena,* 2d ed., Academic, New York, 1963). For most purposes, it is sufficient to view Γ_i as the concentration of adsorbed component i at the surface in units of, say (g·mol)/cm^2. **R** is the gas constant, T is the absolute temperature, γ is the surface tension, and a_i is the activity of component i. The minus sign shows that material which concentrates at the surface generally lowers the surface tension, and vice versa. This can sometimes be a guide in determining preliminarily what materials can be separated.

When applied to a nonionic surfactant in pure water at concentrations below the critical micelle concentration, Eq. (20-39) simplifies into Eq. (20-40)

$$\Gamma_s = -\frac{1}{\mathbf{R}T}\frac{d\gamma}{d\ln C_s} \qquad (20\text{-}40)$$

C is the concentration in the bulk, and subscript s refers to the surfactant. Under some conditions, Eq. (20-40) may apply to an ionic surfactant as well (Lemlich, loc. cit.).

The major surfactant in the foam may usually be considered to be present at the bubble surfaces in the form of an adsorbed monolayer with a substantially constant Γ_s, often of the order of 3×10^{-10} (g·mol)/cm^2, for a molecular weight of several hundred. On the other hand, trace materials follow the linear-adsorption isotherm $\Gamma_i = K_iC_i$ if their concentration is low enough. For a wider range of concentration a Langmuir or other type of isotherm may be applicable (Davies and Rideal, loc. cit.).

Factors Affecting Adsorption K_i for a colligend can be adversely affected (reduced) through an insufficiency of collector. It can also be reduced through an excess of collector, which competes for the available surface against the collector-colligend complex [Schnepf, Gaden, Mirocznik, and Schonfeld, *Chem. Eng. Prog.*, **55**(5), 42 (1959)].

Excess collector can also reduce the separation by forming micelles in the bulk which adsorb some of the colligend, thus keeping it from the surface. This effect of the micelles on K_i for the colligend is given theoretically [Lemlich, "Principles of Foam Fractionation," in Perry (ed.), *Progress in Separation and Purification*, vol. 1, Interscience, New York, 1968, chap. 1] by Eq. (20-41) [Lemlich (ed.), *Adsorptive Bubble Separation Techniques*, Academic, New York, 1972] if Γ_s is constant when $C_s > C_{sc}$:

$$\frac{1}{K_2} = \frac{1}{K_1} + \frac{C_s - C_{sc}}{\Gamma_s E} \tag{20-41}$$

K_1 is K_i just below the collector's critical micelle concentration, C_{sc}. K_2 is K_i at some higher collector concentration, C_s. E is the relative effectiveness, in adsorbing colligend, of surface collector versus micellar collector. Generally, $E > 1$. Γ_s is the surface excess of collector. More about each K is available [Lemlich, "Adsubble Methods," in Li (ed.), *Recent Developments in Separation Science*, vol. 1, CRC Press, Cleveland, 1972, pp. 113–127; Jashnani and Lemlich, *Ind. Eng. Chem. Process Des. Dev.*, **12,** 312 (1973)].

The controlling effect of various ions can be expressed in terms of thermodynamic equilibria [Karger and DeVivo, *Sep. Sci.*, **3,** 393 1968)]. Similarities with ion exchange have been noted. The selectivity of counterionic adsorption increases with ionic charge and decreases with hydration number [Jorne and Rubin, *Sep. Sci.*, **4,** 313 (1969); and Kato and Nakamori, *J. Chem. Eng. Japan*, **9,** 378 (1976)]. By analogy with other separation processes, the relative distribution in multicomponent systems can be analyzed in terms of a selectivity coefficient $\alpha_{mn} = \Gamma_m C_n / \Gamma_n C_m$ [Rubin and Jorne, *Ind. Eng. Chem. Fundam.*, **8,** 474 (1969); *J. Colloid Interface Sci.*, **33,** 208 (1970)].

Operation in the Simple Mode If there is no concentration gradient within the liquid pool and if there is no coalescence within the rising foam, then the operation shown by the solid lines of Fig. 20-41 is truly in the simple mode, i.e., a single theoretical stage of separation. Equations (20-42) and (20-43) will then apply to the steady-flow operation.

$$C_Q = C_W + (GS\Gamma_W/Q) \tag{20-42}$$

$$C_W = C_F - (GS\Gamma_W/F) \tag{20-43}$$

C_F, C_W, and C_Q are the concentrations of the substance in question (which may be a colligend or a surfactant) in the feed stream, bottoms stream, and foamate (collapsed foam), respectively. G, F, and Q are the volumetric flow rates of gas, feed, and foamate, respectively. Γ_W is the surface excess in equilibrium with C_W. S is the surface-to-volume ratio for a bubble. For a spherical bubble, $S = 6/d$, where d is the bubble diameter. For variation in bubble sizes, d should be taken as $\Sigma n_i d_i^3 / \Sigma n_i d_i^2$, where n_i is the number of bubbles with diameter d_i in a representative region of foam.

Finding Γ Either Eq. (20-42) or Eq. (20-43) can be used to find the surface excess indirectly from experimental measurements. To assure a close approach to operation as a single theoretical stage, coalescence in the rising foam should be minimized by maintaining a proper gas rate and a low foam height [Brunner and Lemlich, *Ind. Eng. Chem. Fundam.* **2,** 297 (1963)]. These precautions apply particularly with Eq. (20-42).

For laboratory purposes it is sometimes convenient to recycle the foamate directly to the pool in a manner analogous to an equilibrium still. This eliminates the feed and bottoms streams and makes for a more reliable approach to steady-state operation. However, this recycling may not be advisable for colligend measurements in the presence of slowly dissociating collector micelles.

To avoid spurious effects in the laboratory, it is advisable to employ a prehumidified chemically inert gas.

Bubble Sizes Subject to certain errors (de Vries, *Foam Stability*, Rubber-Stichting, Delft, 1957), foam bubble diameters can be measured photographically. Some of these errors can be minimized by taking pains to generate bubbles of fairly uniform size, say, by using a bubbler with identical orifices or by just using a bubbler with a single orifice (gas rate permitting). Otherwise, a correction for planar statistical sampling bias in the foam should be incorporated with actual diameters [de Vries, op. cit.] or truncated diameters [Lemlich, *Chem. Eng. Commun.* **16,** 153 (1982)]. Also, size segregation can reduce mean mural bubble diameter by roughly half the standard deviation [Cheng and Lemlich, *Ind. Eng. Chem. Fundam.* **22,** 105 (1983)]. Bubble diameters can also be measured in the liquid pool, either photographically or indirectly via measurement of the gas flow rate and stroboscopic determination of bubble frequency [Leonard and Lemlich, *Am. Inst. Chem. Eng. J.*, **11,** 25 (1965)].

Bubble sizes at formation generally increase with surface tension and orifice diameter. Prediction of sizes in swarms from multiple orifices is difficult. In aqueous solutions of low surface tension, bubble diameters of the order of 1 mm are common. Bubbles produced by the more complicated techniques of pressure flotation or vacuum flotation are usually smaller, with diameters of the order of 0.1 mm or less.

Enriching and Stripping Unlike truly simple foam fractionation without significant changes in bubble diameter, coalescence in a foam column destroys some bubble surface and so releases adsorbed material to trickle down through the rising foam. This downflow constitutes internal reflux, which enriches the rising foam by countercurrent action. The result is a richer foamate, i.e., higher C_Q than that obtainable from the single theoretical stage of the corresponding simple mode. Significant coalescence is often present in rising foam, but the effect on bubble diameter and enrichment is frequently overlooked.

External reflux can be furnished by returning some of the externally broken foam to the top of the column. The concentrating effect of reflux, even for a substance which saturates the surface, has been verified [Lemlich and Lavi, *Science*, **134,** 191 (1961)].

Introducing the feed into the foam some distance above the pool makes for stripping operation. The resulting countercurrent flow in the foam further purifies the bottoms, i.e., lowers C_W.

Enriching, stripping, and combined operations are shown in Fig. 20-41.

Foam-Column Theory The counterflowing streams within the foam are viewed as consisting *effectively* of a descending stream of interstitial liquid (equal to zero for the simple mode) and an ascending stream of interstitial liquid plus bubble surface. (By considering this ascending surface as analogous to a vapor, the overall operation becomes analogous in a way to distillation *with entrainment*.)

An effective concentration $[\overline{C}]$ in the ascending stream at any level in the column is defined by Eq. (20-44):

$$\overline{C} = C + (GS\Gamma/U) \tag{20-44}$$

where U is the volumetric rate of interstitial liquid upflow, C is the concentration in this ascending liquid at that level, and Γ is the surface excess in equilibrium with C. Any effect of micelles should be included.

For simplicity, U can usually be equated to Q. An effective equilibrium curve can now be plotted from Eq. (20-44) in terms of $\overline{C}$ (or rather $\overline{C}°$) versus C.

Operating lines can be found in the usual way from material balances. The slope of each such line is $\Delta\overline{C}/\Delta C = L/U$, where L is the downflow rate in the particular column section and C is now the concentration in the descending stream.

The number of theoretical stages can then be found in one of the usual ways. Figure 20-42 illustrates a graphical calculation for a stripper.

Alternatively, the number of transfer units (NTU) in the foam based on, say, the ascending stream can be found from Eq. (20-45):

$$\text{NTU} = \int_{\overline{C}_W}^{C_Q} \frac{d\overline{C}}{\overline{C}° - \overline{C}} \tag{20-45}$$

$\overline{C}°$ is related to C by the effective equilibrium curve, and $\overline{C}_W°$ is similarly related to C_W. $\overline{C}$ is related to C by the operating line.

To illustrate this integration analytically, Eq. (20-45) becomes Eq. (20-46) for the case of a stripping column removing a colligend which is subject to the linear-equilibrium isotherm $\Gamma = KC$.

$$\text{NTU} = \frac{F}{GSK - W} \ln \frac{FW + F(GSK - W)C_F/C_W}{GSK(GSK + F - W)} \tag{20-46}$$

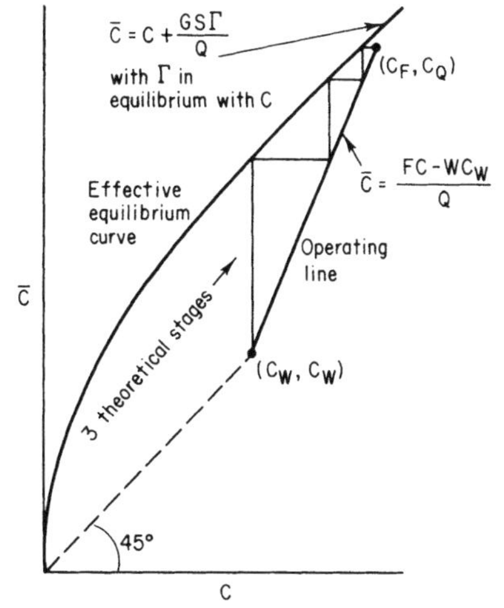

FIG. 20-42 Graphical determination of theoretical stages for a foam-fractionation stripping column.

As another illustration, Eq. (20-45) becomes Eq. (20-47) for an enriching column which is concentrating a surfactant with a constant Γ:

$$\text{NTU} = R \ln \frac{RGS\Gamma(F - D)}{(R + 1)GS\Gamma(F - D) - (R + 1)FD(C_D - C_F)} \tag{20-47}$$

Unless the liquid pool is purposely lengthened vertically in order to give additional separation via bubble fractionation, it is usually taken to represent one theoretical stage. A bubbler submergence of 30 cm or so is usually ample for a solute with a molecular weight that does not exceed several hundred.

In a colligend stripper, it may be necessary to add some collector to the pool as well as the feed because the collector is also stripped off.

Limiting Equations If the height of a foam-fractionation column is increased sufficiently, a concentration pinch will develop between the counterflowing interstitial streams (Brunner and Lemlich, loc. cit.). For an enricher, the separation attained will then approach the predictions of Eq. (20-48) and, interestingly enough, Eq. (20-43).

$$C_D = C_W + (GS\Gamma_W/D) \tag{20-48}$$

D is the volumetric rate at which net foamate (net overhead liquid product) is withdrawn. $D = Q/(R + 1)$. The concentration in the net foamate is C_D. In the usual case of total foam breakage (no dephlegmation), $C_D = C_Q$.

If the tall column is a stripper, the separation will approach that of Eqs. (20-49) and (20-50):

$$C_Q = C_F + (GS\Gamma_F/Q) \tag{20-49}$$

$$C_W = C_F - (GS\Gamma_F/W) \tag{20-50}$$

For a sufficiently tall combined column, the separation will approach that of Eqs. (20-51) and (20-50):

$$C_D = C_F + (GS\Gamma_F/D) \tag{20-51}$$

The formation of micelles in the foam breaker does not affect the limiting equations because of the theoretically unlimited opportunity in a sufficiently tall column for their transfer from the reflux to the ascending stream [Lemlich, "Principles of Foam Fractionation," in Perry (ed.), *Progress in Separation and Purification,* vol. 1, Interscience, New York, 1968, chap. 1].

In practice, the performance of a well-operated foam column several feet tall may actually approximate the limiting equations, provided there is little channeling in the foam and provided that reflux is either absent or is present at a low ratio.

Column Operation To assure intimate contact between the counterflowing interstitial streams, the volume fraction of liquid in the foam should be kept below about 10 percent—and the lower the better. Also, rather uniform bubble sizes are desirable. The foam bubbles will thus pack together as blunted polyhedra rather than as spheres, and the suction in the capillaries (Plateau borders) so formed will promote good liquid distribution and contact. To allow for this desirable deviation from sphericity, $S = 6.3/d$ in the equations for enriching, stripping, and combined column operation [Lemlich, *Chem. Eng.,* **75**(27), 95 (1968); **76**(6), 5 (1969)]. Diameter d still refers to the sphere.

Visible channeling or significant deviations from plug flow of the foam should be avoided, if necessary by widening the column or lowering the gas and/or liquid rates. The superficial gas velocity should probably not exceed 1 or 2 cm/s. Under proper conditions, HTU values of several cm have been reported [Hastings, Ph.D. dissertation, Michigan State University, East Lansing, 1967; and Jashnani and Lemlich, *Ind. Eng. Chem. Process Des. Dev.,* **12,** 312 (1973)]. The foam column height equals NTU × HTU.

For columns that are wider than several centimeters, reflux and feed distributors should be used, particularly for wet foam [Haas and Johnson, *Am. Inst. Chem. Eng. J.,* **11,** 319 (1965)]. Liquid content within the foam can be monitored conductometrically [Chang and Lemlich, *J. Colloid Interface Sci.,* **73,** 224 (1980)]. See Fig. 20-43. Theoretically, as the limit $\mathcal{D} = K = 0$ is very closely approached, $\mathcal{D} = 3K$ [Lemlich, *J. Colloid Interface Sci.,* **64,** 107 (1978)].

Wet foam can be handled in a bubble-cap column (Wace and Banfield, *Chem. Process Eng.,* **47**(10), 70 (1966)] or in a sieve plate column [Aguayo and Lemlich, *Ind. Eng. Chem. Process Des. Dev.,* **13,** 153 (1974)]. Alternatively, individual short columns can be connected in countercurrent array [Banfield, Newson, and Alder, *Am. Inst. Chem. Eng. Symp. Ser.,* **1,** 3 (1965); Leonard and Blacyki, *Ind. Eng. Chem. Process Des. Dev.,* **17,** 358 (1978)].

A high gas rate can be used to achieve maximum throughput in the simple mode (Wace, Alder, and Banfield, AERE-R5920, U.K. Atomic Energy Authority, 1968) because channeling is not a factor in that mode. A horizontal drainage section can be used overhead [Haas and Johnson, *Ind. Eng. Chem. Fundam.,* **6,** 225 (1967)]. The highly mobile dispersion produced by a very high gas rate is not a true foam but is

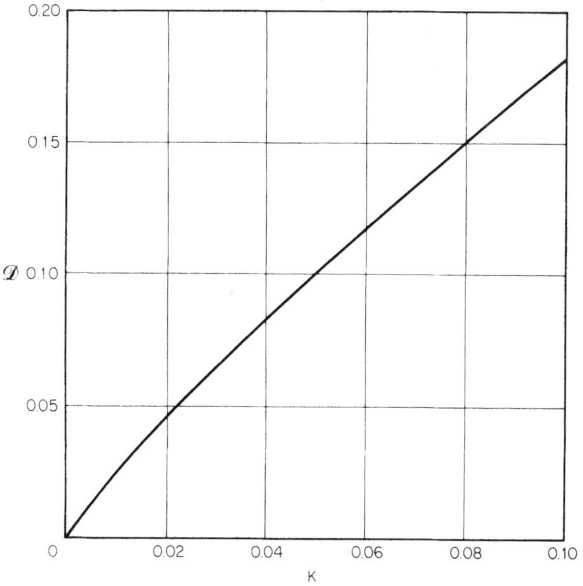

FIG. 20-43 Empirical relationship between $\mathcal{D}$, the volumetric fraction of liquid in common polydisperse foam, and K, the electrical conductivity of the foam divided by the electrical conductivity of the liquid. [*Chang and Lemlich,* J. Colloid Interface Sci., ***73,** 224 (1980).*]

rather a so-called **gas emulsion** [Bikerman, *Ind. Eng. Chem.*, **57**(1), 56 (1965)].

A very low gas rate in a column several feet tall with internal reflux can sometimes be used to effect difficult multicomponent separations in batch operation [Lemlich, "Principles of Foam Fractionation," in Perry (ed.), *Progress in Separation and Purification*, vol. 1, Interscience, New York, 1968, chap. 1].

The same end may be achieved by continuous operation at total external reflux with a small U bend in the reflux line for foamate holdup [Rubin and Melech, *Can. J. Chem. Eng.*, **50,** 748 (1972)].

The slowly rising foam in a tall column can be employed as the sorbent for continuous chromatographic separations [Talman and Rubin, *Sep. Sci.*, **11,** 509 (1976)]. Low gas rates are also employed in short columns to produce the scumlike froth of batch-operated ion flotation, microflotation, and precipitate flotation.

Foam Drainage and Overflow The rate of foam overflow on a gas-free basis (i.e., the total volumetric foamate rate Q) can be estimated from a detailed theory for foam drainage [Leonard and Lemlich, *Am. Inst. Chem. Eng. J.*, **11,** 18 (1965)]. From the resulting relationship for overflow [Fanlo and Lemlich, *Am. Inst. Chem. Eng. Symp. Ser.*, **9,** 75, 85 (1965)], Eq. (20-52) can be employed as a convenient approximation to the theory so as to avoid trial and error over the usual range of interest for foam of low liquid content ascending in plug flow:

$$\frac{Q}{G} = 22\left(\frac{v_G^3 \mu \mu_s^2}{g^3 \rho^3 d^8}\right)^{1/4} \tag{20-52}$$

The superficial gas velocity v_g is G/A, where A is the horizontal cross-sectional area of the empty vertical foam column. Also, g is the acceleration of gravity, ρ is the liquid density, μ is the ordinary liquid viscosity, and μ_s is the effective surface viscosity.

To account for inhomogeneity in bubble sizes, d in Eq. (20-52) should be taken as $\sqrt{\Sigma n_i d_i^3/\Sigma n_i d_i}$ and evaluated at the top of the vertical column if coalescence is significant in the rising foam. Note that this average d for overflow differs from that employed earlier for S. Also, see "Bubble Sizes" regarding the correction for planar statistical sampling bias and the presence of size segregation at a wall.

For theoretical reasons, Q determined from Eq. (20-52) should be multiplied by the factor $(1 + 3Q/G)$ to give a final Q. However, for foam of sufficiently low liquid content this multiplication can be omitted with little error.

The effective surface viscosity is best found by experiment with the system in question, followed by back calculation through Eq. (20-52). From the precursors to Eq. (20-52), such experiments have yielded values of μ_s on the order of 10^{-4} (dyn·s)/cm for common surfactants in water at room temperature, which agrees with independent measurements [Lemlich, *Chem. Eng. Sci.*, **23,** 932 (1968); and Shih and Lemlich, *Am. Inst. Chem. Eng. J.*, **13,** 751 (1967)]. However, the expected high μ_s for aqueous solutions of such skin-forming substances as saponin and albumin was not attained, perhaps because of their non-newtonian surface behavior [Shih and Lemlich, *Ind. Eng. Chem. Fundam.*, **10,** 254 (1971); and Jashnani and Lemlich, *J. Colloid Interface Sci.*, **46,** 13 (1974)].

The drainage theory breaks down for columns with tortuous cross section, large slugs of gas, or heavy coalescence in the rising foam.

Foam Coalescence Coalescence is of two types. The first is the growth of the larger foam bubbles at the expense of the smaller bubbles due to interbubble gas diffusion, which results from the smaller bubbles having somewhat higher internal pressures (Adamson and Gast, *Physical Chemistry of Surfaces*, 6th ed., Wiley, New York, 1997). Small bubbles can even disappear entirely. In principle, the rate at which this type of coalescence proceeds can be estimated [Ranadive and Lemlich, *J. Colloid Interface Sci.*, **70,** 392 (1979)].

The second type of coalescence arises from the rupture of films between adjacent bubbles [Vrij and Overbeek, *J. Am. Chem. Soc.*, **90,** 3074 (1968)]. Its rate appears to follow first-order reaction kinetics with respect to the number of bubbles [New, *Proc. 4th Int. Congr. Surf. Active Substances*, Brussels, 1964, **2,** 1167 (1967)] and to decrease with film thickness [Steiner, Hunkeler, and Hartland, *Trans. Inst. Chem. Eng.*, **55,** 153 (1977)]. Many factors are involved [Bikerman, *Foams*, Springer-Verlag, New York, 1973; and Akers (ed.), *Foams*, Academic, New York, 1976].

Both types of coalescence can be important in the foam separations characterized by low gas flow rate, such as batchwise ion flotation producing a scum-bearing froth of comparatively long residence time. On the other hand, with the relatively higher gas flow rate of foam fractionation, the residence time may be too short for the first type to be important, and *if* the foam is sufficiently stable, even the second type of coalescence may be unimportant.

Unlike the case for Eq. (20-52), when coalescence is significant, it is better to find S from d evaluated at the feed level for Eqs. (20-49) to (20-51) and at the pool surface for Eqs. (20-43) and (20-48).

Foam Breaking It is usually desirable to collapse the overflowing foam. This can be accomplished by chemical means (Bikerman, op. cit.) if external reflux is not employed or by thermal means [Kishimoto, *Kolloid Z.*, **192,** 66 (1963)] if degradation of the overhead product is not a factor.

Foam can also be broken with a rotating perforated basket [Lemlich, "Principles of Foam Fractionation," in Perry (ed.), *Progress in Separation and Purification*, vol. 1, Interscience, New York, 1968, chap. 1]. If the foamate is aqueous (as it usually is), the operation can be improved by discharging onto Teflon instead of glass [Haas and Johnson, *Am. Inst. Chem. Eng. J.*, **11,** 319 (1965)]. A turbine can be used to break foam [Ng, Mueller, and Walden, *Can. J. Chem. Eng.*, **55,** 439 (1977)]. Foam which is not overly stable has been broken by running foamate onto it [Brunner and Stephan, *Ind. Eng. Chem.*, **57**(5), 40 (1965)]. Foam can also be broken by sound or ultrasound, a rotating disk, and other means [Ohkawa, Sakagama, Sakai, Futai, and Takahara, *J. Ferment. Technol.*, **56,** 428, 532 (1978)].

If desired, dephlegmation (partial collapse of the foam to give reflux) can be accomplished by simply widening the top of the column, provided the foam is not too stable. Otherwise, one of the more positive methods of foam breaking can be employed to achieve dephlegmation.

Bubble Fractionation Figure 20-44 shows continuous bubble fractionation. This operation can be analyzed in a simplified way in terms of the adsorbed carry-up, which furthers the concentration gradient, and the dispersion in the liquid, which reduces the gradient [Lemlich, *Am. Inst. Chem. Eng. J.*, **12,** 802 (1966); **13,** 1017 (1967)].

To illustrate, consider the limiting case in which the feed stream and the two liquid takeoff streams of Fig. 20-44 are each zero, thus resulting in batch operation. At steady state the rate of adsorbed carry-up will equal the rate of downward dispersion, or $af\Gamma = \overline{D}AdC/dh$. Here a is the surface area of a bubble, f is the frequency of bubble formation. $\overline{D}$ is the dispersion (effective diffusion) coefficient based on the column cross-sectional area A, and C is the concentration at height h within the column.

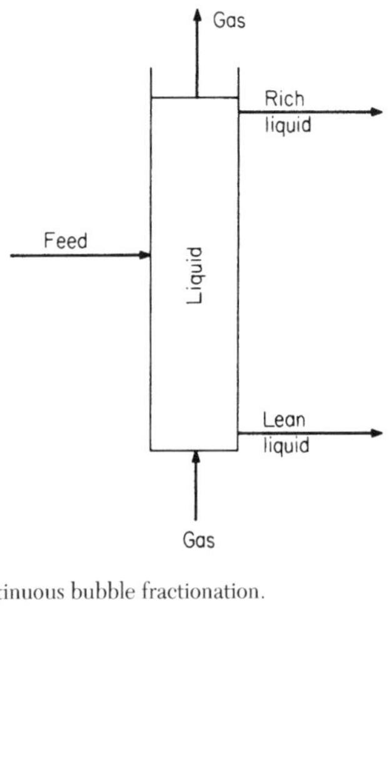

FIG. 20-44 Continuous bubble fractionation.

There are several possible alternative relationships for Γ (Lemlich, op. cit.). For simplicity, consider $\Gamma = K'C$, where K' is not necessarily the same as the equilibrium constant K. Substituting and integrating from the boundary condition of $C = C_B$ at $h = 0$ yield

$$C/C_B = \exp(Jh) \tag{20-53}$$

C_B is the concentration at the bottom of the column, and parameter $J = K'af/\overline{D}A$. Combining Eq. (20-53) with a material balance against the solute in the initial charge of liquid gives

$$\frac{C}{C_i} = \frac{JH \exp(Jh)}{\exp(JH) - 1} \tag{20-54}$$

C_i is the concentration in the initial charge, and H is the total height of the column.

The foregoing approach has been extended to steady continuous flow as illustrated in Fig. 20-44 [Cannon and Lemlich, *Chem. Eng. Prog. Symp. Ser.*, **68**(124), 180 (1972); Bruin, Hudson, and Morgan, *Ind. Eng. Chem. Fundam.*, **11**, 175 (1972); and Wang, Granstrom, and Kown, *Environ. Lett.*, **3**, 251 (1972), **4**, 233 (1973), **5**, 71 (1973)]. The extension includes a rough method for estimating the optimum feed location as well as a very detailed analysis of column performance which takes into account the various local phenomena around each rising bubble (Cannon and Lemlich, op. cit.).

Uraizee and Narsimhan [*Sep. Sci. Technol.*, **30**(6), 847 (1995)] have provided a model for the continuous separation of proteins from dilute solutions. Although their work is focused on protein separation, the model should find general application to other separations.

In agreement with experiment [Shah and Lemlich, *Ind. Eng. Chem. Fundam.*, **9**, 350 (1970); and Garmendia, Perez, and Katz, *J. Chem. Educ.*, **50**, 864 (1973)], theory shows that the degree of separation that is obtained increases as the liquid column is made taller. But unfortunately it decreases as the column is made wider. In simple terms, the latter effect can be attributed to the increase in the dispersion coefficient as the column is widened.

In this last connection it is important that the column be aligned precisely vertically (Valdes-Krieg, King, and Sephton, *Am. Inst. Chem. Eng. J.*, **21**, 400 (1975)]. Otherwise, the bubbles with their dragged liquid will tend to rise up one side of the column, thus causing liquid to flow down the other side, and in this way largely destroy the concentration gradient. A vertical foam-fractionation column should also be carefully aligned to be plumb.

The escaping bubbles from the top of a bubble-fractionation column can carry off an appreciable quantity of adsorbed material in an aerosol of very fine film drops [various papers, *J. Geophys. Res., Oceans Atmos.*, **77**(27), (1972)]. If the residual solute is thus appreciably depleted, C_i in Eq. (20-54) should be replaced with the average residual concentration.

This carry-off of film drops, which may also occur with breaking foam, in certain cases can partially convert water pollution into air pollution. If such is the case, it may be desirable to recirculate the gas. Such recirculation is also indicated if hydrocarbon vapors or other volatiles are incorporated in the gas stream to improve adsorptive selectivity [Maas, *Sep. Sci.*, **4**, 457 (1969)].

A small amount of collector (surfactant) or other appropriate additive in the liquid may greatly increase adsorption (Shah and Lemlich, op. cit.). Column performance can also be improved by skimming the surface of the liquid pool or, when possible, by removing adsorbed solute in even a tenuous foam overflow. Alternatively, an immiscible liquid can be floated on top. Then the concentration gradient in the tall pool of main liquid, plus the trapping action of the immiscible layer above it, will yield a combination of bubble fractionation and solvent sublation.

Systems Separated Some of the various separations reported in the literature are listed in Rubin and Gaden, "Foam Separation," in Schoen (ed.), *New Chemical Engineering Separation Techniques*, Interscience, New York, 1962, chap. 5; Lemlich, *Ind. Eng. Chem.*, **60**(10), 16 (1968); Pushkarev, Egorov, and Khrustalev, *Clarification and Deactivation of Waste Waters by Frothing Flotation*, in Russian, Atomizdat, Moscow, 1969; Kuskin and Golman, *Flotation of Ions and Molecules*, in Russian, Nedra, Moscow, 1971; Lemlich (ed.), *Adsorptive Bubble Separation Techniques*, Academic, New York, 1972; Lemlich, "Adsubble Methods," in Li (ed.), *Recent Developments in Separation Science*, vol. 1, CRC Press, Cleveland, 1972, chap. 5; Grieves, *Chem. Eng. J.*, **9**, 93 (1975); Valdes-Krieg King, and Sephton, *Sep. Purif. Methods*, **6**, 221 (1977); Clarke and Wilson, *Foam Flotation*, Marcel Dekker, New York, 1983; and Wilson and Clarke, "Bubble and Foam Separations in Waste Treatment," in Rousseau (ed.), *Handbook of Separation Processes*, Wiley, New York, 1987.

Of the numerous separations reported, only a few can be listed here. Except for minerals beneficiation [ore flotation], the most important industrial applications are usually in the area of pollution control.

A pilot-sized foaming unit reduced the alkyl benzene sulfonate concentration of 500,000 gal of sewage per day to nearly 1 mg/L, using a G/F of 5 and producing a Q/F of no more than 0.03 [Brunner and Stephan, *Ind. Eng. Chem.*, **57**(5), 40 (1965); and Stephan, *Civ. Eng.*, **35**(9), 46 (1965)]. A full-scale unit handling over 45,420 m^3/day (12 million gal/day) performed nearly as well. The foam also carried off some other pollutants. However, with the widespread advent of biodegradable detergents, large-scale foam fractionation of municipal sewage has been discontinued.

Other plant-scale applications to pollution control include the flotation of suspended sewage particles by depressurizing so as to release dissolved air [Jenkins, Scherfig, and Eckhoff, "Applications of Adsorptive Bubble Separation Techniques to Wastewater Treatment," in Lemlich (ed.), *Adsorptive Bubble Separation Techniques*, Academic, New York, 1972, chap. 14; and Richter, *Internat. Chem. Eng.*, **16**, 614 (1976)]. Dissolved-air flotation is also employed in treating wastewater from pulp and paper mills [Coertze, *Prog. Water Technol.*, **10**, 449 (1978); and Severeid, *TAPPI* **62**(2), 61, 1979]. In addition, there is the flotation, with electrolytically released bubbles [Chambers and Cottrell, *Chem. Eng.*, **83**(16), 95 (1976)], of oily iron dust [Ellwood, *Chem. Eng.*, **75**(16), 82 (1968)] and of a variety of wastes from surface-treatment processes at the maintenance and overhaul base of an airline [Roth and Ferguson, *Desalination*, **23**, 49 (1977)].

Fats and, through the use of lignosulfonic acid, proteins can be flotated from the wastewaters of slaughterhouses and other food-processing installations [Hopwood, *Inst. Chem. Eng. Symp. Ser.*, **41**, M1 (1975)]. After further treatment, the floated sludge has been fed to swine.

A report of the recovery of protein from potato-juice wastewater by foaming [Weijenberg, Mulder, Drinkenberg, and Stemerding, *Ind. Eng. Chem. Process Des. Dev.*, **17**, 209 (1978)] is reminiscent of the classical recovery of protein from potato and sugar-beet juices [Ostwald and Siehr, *Kolloid Z.*, **79**, 11 (1937)]. The isoelectric pH is often a good choice for the foam fractionation of protein (Rubin and Gaden, loc. cit.). Adding a salt to lower solubility may also help. Additional applications of foam fractionation to the separation of protein have been reported by Uraizee and Narsimhan [*Enzyme Microb. Technol.* **12**, 232 (1990)].

With the addition of appropriate additives as needed, the flotation of refinery wastewaters reduced their oil content to less than 10 mg/L in pilot-plant operation [Steiner, Bennett, Mohler, and Clere, *Chem. Eng. Prog.*, **74**(12), 39 (1978)] and full-scale operation (Simonsen, *Hydrocarb. Process. Pet. Refiner*, **41**(5), 145, 1962]. Experiments with a cationic collector to remove oils reportedly confirmed theory [Angelidon, Keskavarz, Richardson, and Jameson, *Ind. Eng. Chem. Process Des. Dev.*, **16**, 436 (1977)].

Pilot-plant [Hyde, Miller, Packham, and Richards, *J. Am. Water Works Assoc.*, **69**, 369 (1977)] and full-scale [Ward, *Water Serv.*, **81**, 499 (1977)] flotation in the preparation of potable water is described.

Overflow at the rate of 2700 m^3 (713,000 gal) per day from a zinc-concentrate thickener is treated by ion flotation, precipitate flotation, and untrafine-particle flotation [Nagahama, *Can. Min. Metall. Bull.*, **67**, 79 (1974)]. In precipitate flotation only the surface of the particles need be coated with collector. Therefore, in principle less collector is required than for the equivalent removal of ions by foam fractionation or ion flotation.

By using an anionic collector and external reflux in a combined (enriching and stripping) column of 3.8-cm (1.5-in) diameter with a feed rate of 1.63 m/h [40 gal/(h·ft^2)] based on column cross section,

D/F was reduced to 0.00027 with C_w/C_F for Sr^{2+} below 0.001 [Shonfeld and Kibbey, *Nucl. Appl.*, **3,** 353 (1967)]. Reports of the adsubble separation of 29 heavy metals, radioactive and otherwise, have been tabulated [Lemlich, "The Adsorptive Bubble Separation Techniques," in Sabadell (ed.), *Proc. Conf. Traces Heavy Met. Water,* 211–223, Princeton University, 1973, EPA 902/9-74-001, U.S. EPA, Reg. II, 1974). Some separation of ^{15}N from ^{14}N by foam fractionation has been reported [Hitchcock, Ph.D. dissertation, University of Missouri, Rolla, 1982].

The numerous separations reported in the literature include surfactants, inorganic ions, enzymes, other proteins, other organics, biological cells, and various other particles and substances. The scale of the systems ranges from the simple Crits test for the presence of surfactants in water, which has been shown to operate by virtue of transient foam fractionation [Lemlich, *J. Colloid Interface Sci.*, **37,** 497 (1971)], to the natural adsubble processes that occur on a grand scale in the ocean [Wallace and Duce, *Deep Sea Res.*, **25,** 827 (1978)]. For further information see the reviews cited earlier.

MEMBRANE SEPARATION PROCESSES

General References: Cheryan, *Ultrafiltration and Microfiltration Handbook*, Technomic Publishing Company, Pa., 1988. Ho and Sirkar (eds.), *Membrane Handbook*, Van Nostrand Reinhold, New York, 1992. Baker, *Membrane Technology and Applications,* 2d ed., Wiley, 2000. Eykamp, Sec. 22: Membrane Separation Processes, in *Perry's Chemical Engineers' Handbook,* 7th ed., Perry and Green (eds.), McGraw-Hill, New York, 1997. Millipore Corporation, *Protein Concentration and Diafiltration by Tangential Flow Filtration*, Lit. No. TB032 Rev. B, 1999. Zeman and Zydney, *Microfiltration and Ultrafiltration: Principles and Applications*, Marcel Dekker, New York, 1996. A review of membrane retention mechanisms can be found in Deen, *AIChE J.*, **33,** 1409–1425 (1987).

New developments in membranes are found in journals and trade magazines (e.g., *J. Membrane Sci., BioPharm International*), vendor communications (e.g., websites), patent filings, and conference presentations (e.g., annual ACS or NAMS meetings). Areas of active research include new membrane polymers and surface modification with accompanying diagnostic methods (to reduce fouling, increase flux and retention, improve consistency), new module designs (to improve flux, cleanability, ease of use, scalability, reliability), new processing skids (better components, recovery, less holdup, better mixing, disposability, software for automated processing and archiving), new processing methods (diafiltration strategies, turbulence enhancements), and new applications (e.g., protein-protein separations, plasmids).

Topics Omitted from This Section In order to concentrate on the membrane processes of widest industrial interest, several are left out.

Dialysis and Hemodialysis Historically, dialysis has found some industrial use. Today, much of that is supplanted by ultrafiltration. Donan dialysis is treated briefly under electrodialysis. Hemodialysis is a huge application for membranes, and it dominates the membrane field in area produced and in monetary value. This medical application is omitted here.

An excellent description of the engineering side of both topics is provided by Kessler and Klein [in Ho and Sirkar (eds.), op. cit., pp. 163–216]. A comprehensive treatment of diffusion appears in: Von Halle and Shachter, "Diffusional Separation Methods," in *Encyclopedia of Chemical Technology,* pp. 149–203, Wiley, 1993.

Facilitated Transport Transport by a reactive phase through a membrane is promising but problematic. Way and Noble [in Ho and Sirkar (eds.), op. cit., pp. 833–866] have a description and a complete bibliography.

Liquid Membranes Several types of liquid membranes exist: molten salt, emulsion, immobilized/supported, and hollow-fiber-contained liquid membranes. Araki and Tsukube (*Liquid Membranes: Chemical Applications,* CRC Press, 1990) and Sec. IX and Chap. 42 in Ho and Sirkar (eds.) (op. cit., pp. 724, 764–808) contain detailed information and extensive bibliographies.

Catalytic Membranes Falconer, Noble, and Sperry (Chap. 14—"Catalytic Membrane Reactors" in Noble and Stern, op. cit., p. 669–712) give a detailed review and an extensive bibliography. Additional information can be found in a work by Tsotsis et al. ["Catalytic Membrane Reactors," pp. 471–551, in Becker and Pereira (eds.), *Computer-Aided Design of Catalysts,* Dekker, 1993].

GENERAL BACKGROUND AND DEFINITIONS

Applications Membranes create a boundary between different bulk gas or liquid mixtures. Different solutes and solvents flow through membranes at different rates. This enables the use of membranes in separation processes. Membrane processes can be operated at moderate temperatures for sensitive components (e.g., food, pharmaceuticals). Membrane processes also tend to have low relative capital and energy costs. Their modular format permits reliable scale-up and operation. This unit operation has seen widespread commercial adoption since the 1960s for component enrichment, depletion, or equilibration. Estimates of annual membrane module sales in 2005 are shown in Table 20-16. Applications of membranes for diagnostic and bench-scale use are not included. Natural biological systems widely employ membranes to isolate cells, organs, and nuclei.

Common Definitions Membrane processes have been evolving since the 1960s with each application tending to generate its own terminology. Recommended nomenclature is provided along with alternatives in current use.

Fluid Stream Designations For the generalized membrane module shown in Fig. 20-45, a feed stream enters a membrane module while both a permeate and a retentate stream exit the module. The permeate (or filtrate) stream flows through the membrane and has been depleted of retained components. The term *filtrate* is commonly used for NFF operation while permeate is used for TFF operation. The retentate (or concentrate) stream flows through the module, not the membrane, and has been enriched in retained components.

TABLE 20-16 Membrane Market in 2005

Segment	$M/yr Size	Applications	Characteristics
Dialysis	~2,000	Medical	Mature growing 5%
Reverse osmosis	~500	Water treatment	Growing 10%
Microfiltration	~500	Water, food, pharm.	
Ultrafiltration	~400	Water, food, pharm.	Growing 10%
Gas separation	~500	Nitrogen	
Electrodialysis	~100	Water	
Pervaporation	~5	Solvent/water	Nascent
Facilitated transport	0	None	In development

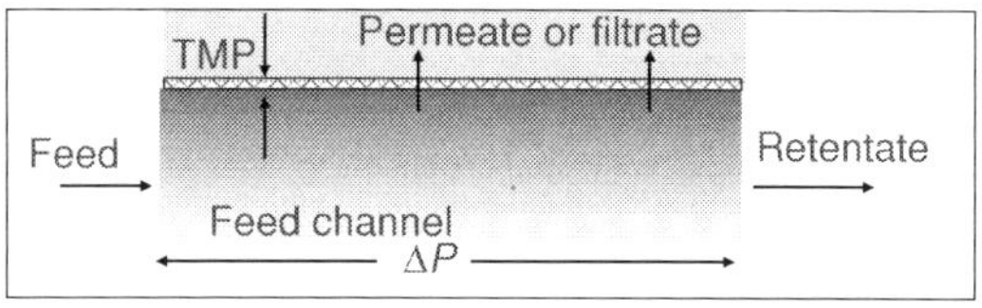

FIG. 20-45 Fluid stream schematic.

Flow Types: Normal Flow and Tangential Flow Filtration If the retentate flow in Fig. 20-45 is zero and all the feed stream flows as a velocity vector normal to the membrane surface, this type of filtration is referred to as *normal flow filtration* (*NFF*), also called dead-end flow. If there is a stream that flows as a velocity vector tangent to the membrane surface, creating a velocity gradient at the membrane surface, and that exits the module as a retentate stream, this is referred to as *tangential flow filtration* (*TFF*), also called crossflow filtration. A retentate stream that flows through a module without creating a surface velocity gradient is merely a bypass and not TFF.

TFF mitigates the accumulation of retained components on the membrane surface, reducing the plugging of the membrane and permitting a more steady-state operation. TFF is used in processing feeds with a high concentration of retained components.

Flow: Flux, Permeability, Conversion The productivity of a membrane module is described by its flux J = volumetric permeate flow rate/membrane area with units of volume per area per time. Relatively high flux rates imply that relatively small membrane areas are required. The permeate volume is usually greater than the feed volume for a given process. Flux is also the magnitude of the normal flow velocity with units of distance per time.

The sensitivity of productivity or flux to *transmembrane pressure* (*TMP*) is referred to as the permeability L = flux/transmembrane pressure. TMP refers to a module average. Pure-component permeability (e.g., water permeability) refers to membrane properties while the more industrially relevant process permeability includes fouling and polarization effects.

The efficiency of a membrane module is characterized by the recovery or conversion ratio CR = permeate flow rate/feed flow rate. Low conversion means that fluid has to be repeatedly cycled past the TFF module to generate permeate. High-efficiency NFF has CR = 1.

Flux Decline: Plugging, Fouling, Polarization Membranes operated in NFF mode tend to show a steady flux decline while those operated in TFF mode tend to show a more stable flux after a short initial decline. Irreversible flux decline can occur by membrane compression or retentate channel spacers blinding off the membrane. Flux decline by fouling mechanisms (molecular adsorption, precipitation on the membrane surface, entrapment within the membrane structure) are amenable to chemical cleaning between batches. Flux decline amenable to mechanical disturbance (such as TFF operation) includes the formation of a secondary structure on the membrane surface such as a static cake or a fluid region of high component concentration called a polarization layer.

Understanding polarization and controlling its effects are key to implementing a good TFF process. Solutes entrained by the permeate flow are retained by the membrane. They accumulate on the membrane surface and form a region of high concentration called the polarization boundary layer. A steady state is reached between back transport away from the membrane surface, tangential convective transport along the membrane surface, and normal convective flow toward the membrane. The back transport leading to steady-state operation gives TFF a high capacity. Plugging is commonly used to describe flux decline through fouling and caking mechanisms.

Single-Component Separation: Passage, Retention, LRV The passage of a component through a membrane (also called the sieving coefficient or transmission) is calculated as the ratio $S = c_P/c_F$, where c_P and c_F refer to the permeate concentration, and the feed concentration, respectively. These concentrations may change with position within a module. In industrial practice, it is common to measure these concentrations in the permeate stream exiting a module and the feed stream entering a module to give an observed passage or S_{obs}.

Membrane vendors and researchers may use the feed concentration at the membrane surface or c_w to calculate an intrinsic passage S_{int}, $= c_P/c_w$. The intrinsic passage characterizes the membrane while the observed passage characterizes the module.

Complementary to the passage, one can also consider the *retention* of a component as $R = 1 - S$ (also called rejection). Retention can also be either an observed or an intrinsic measurement. Retention is useful in considering retained products during concentration mode operation. Other component separation characterizations include the *log reduction value* LRV $= -\log S$ which is used to characterize high-efficiency separations with permeate products (sterilization). The beta ratio $\beta = 1/S$ is sometimes used in NFF for clarification applications.

Multiple-Component Separation: Separation Factor Consistent with the characterization of different separation methods, one can define a separation factor α_{ij} (also called selectivity) for components i and j that compares their relative concentrations in the permeate stream to those in the feed stream:

$$\alpha_{ij} = \frac{c_{iP}/c_{jP}}{c_{iF}/c_{jF}} = \frac{S_i}{S_j} \tag{20-55}$$

The convention is to designate the component with higher passage as component i so that $\alpha_{ij} > 1$.

Membrane Types Key membrane properties include their size rating, selectivity, permeability, mechanical robustness (to allow module fabrication and withstand operating conditions), chemical robustness (to fabrication materials, process fluids, cleaners, and sanitizers), low extractibles, low fouling characteristics, high capacity, low cost, and consistency.

Size Ratings The relative sizes of common components and the associated membrane classes capable of retaining them are shown in Fig. 20-46.

Vendors characterize their filters with ratings indicating the approximate size (or corresponding molecular weight) of components retained by the membrane. This rating should be used as a rough guide only and followed up with retention testing. Among the factors affecting retention are the application-specific retention requirements, variable component size and shapes depending on solution environment, membrane fouling and compaction, degree of module polarization, and interaction between feed components.

Composition and Structure Commercial membranes consist primarily of polymers and some ceramics. Other membrane types include sintered metal, glass, and liquid film. Polymeric membranes are formed by precipitating a 5 to 25 wt % casting solution (lacquer) into a film by solvent evaporation (air casting), liquid extraction (immersion casting), or cooling (melt casting or thermally induced phase separation). Membranes can be cast as flat sheets on a variety of supports or as fibers through a die. Ceramic membranes are formed by depositing successive layers of smaller and smaller inorganic particles on a monolith substrate, to create smaller and smaller interstices or "pores" between particles. These layers are then sintered below the melting point to create a rigid structure.

Membranes may be surface-modified to reduce fouling or improve chemical resistance. This can involve adding surface-modifying agents directly to the lacquer or modifying the cast membrane through chemical or physical treatment. Membranes can also be formed by selective etching or track-etching (radiation treatment followed by etching). Stretching is used to change pore morphology.

Liquid film membranes consist of immiscible solutions held in membrane supports by capillary forces. The chemical composition of these solutions is designed to enhance transport rates of selected components through them by solubility or coupled chemical reaction.

Membrane Morphology—Pores, Symmetric, Composite Only nucleopore and anodyne membranes have relatively uniform pores. Reverse osmosis, gas permeation, and pervaporation membranes have nonuniform angstrom-sized pores corresponding to spaces in between the rigid or dynamic membrane molecules. Solute-membrane molecular interactions are very high. Ultrafiltration membranes have nonuniform nanometer sized pores with some solute-membrane interactions. For other microfiltration membranes with nonuniform pores on the submicrometer to micrometer range, solute-membrane interactions are small.

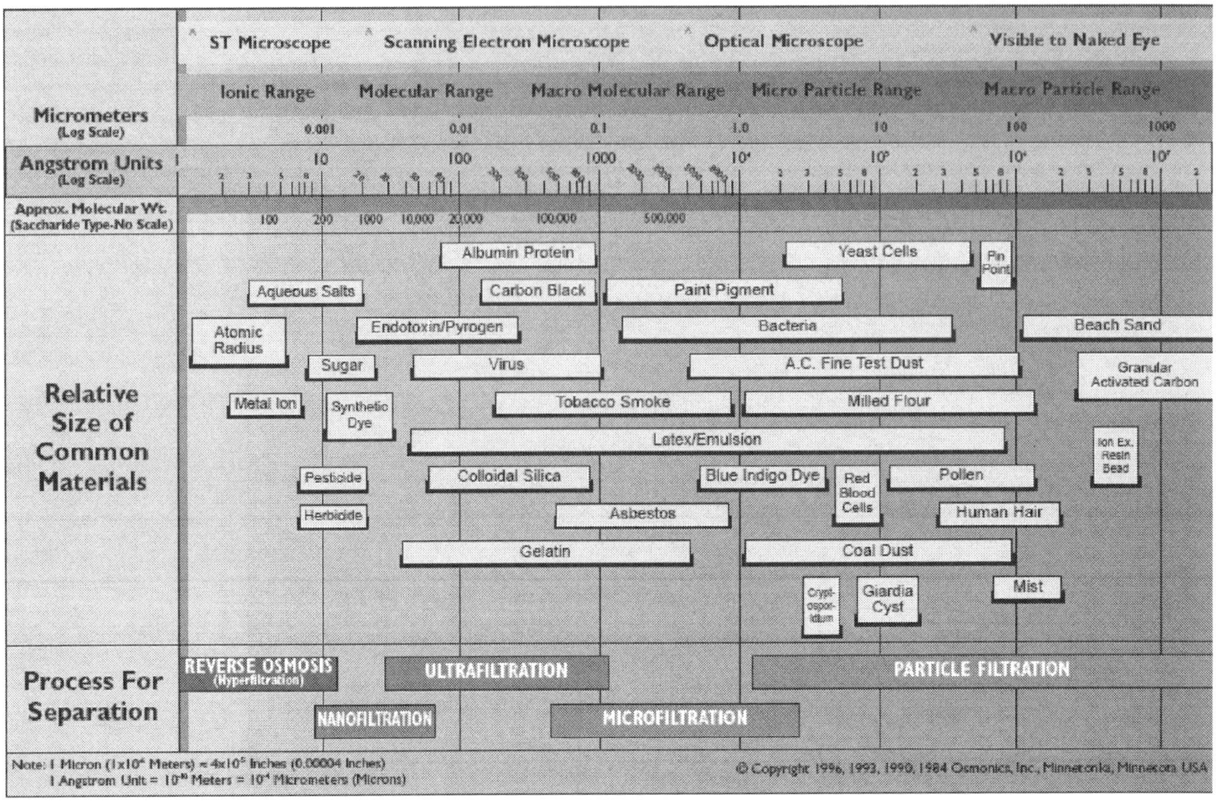

FIG. 20-46 Size spectrum. (*Copyright © 1996. Reprinted by permission of Osmonics, Minnetonka, Minn.*)

Membranes with a relatively uniform pore size distribution throughout the thickness of the membrane are referred to as symmetric or homogeneous membranes. Others may be formed with tight skin layers on the top or on both the top and bottom of the membrane surfaces. These are referred to as asymmetric or nonhomogeneous membranes. In addition, membranes can be cast on top of each other to form a composite membrane.

Asymmetric membranes have a "tight," low-permeability, retentive zone that performs the desired separation and a more open, high-permeability zone that provides mechanical strength to the overall membrane. This structure is particularly critical to the economic viability of reverse-osmosis membranes. Asymmetric membranes operated in TFF mode must have the tight side facing the feed channel so that particles are retained on its surface and can be acted upon by the tangential flow. Asymmetric membranes operated in NFF mode can be operated with the open side facing the feed flow so that retained particles can penetrate the membrane and be dispersed without blinding off the membrane surface and plugging it quickly.

Component Transport Transport through membranes can be considered as mass transfer in series: (1) transport through a polarization layer above the membrane that may include static or dynamic cake layers, (2) partitioning between the upstream polarization layer and membrane phases at the membrane surface, (3) transport through the membrane, and (4) partitioning between the membrane and downstream fluid.

Surface Polarization in TFF The simplified model of polarization shown in Fig. 20-47 is used as a basis for analyzing more complex systems. Consider a single component with no reaction in a thin, two-dimensional boundary layer near the membrane surface. Axial diffusion is negligible along the membrane surface compared to convection.

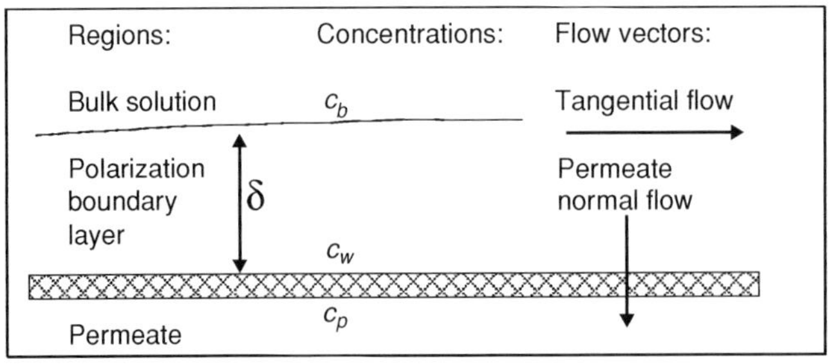

FIG. 20-47 Polarization in tangential flow filtration.

Fluid density and component brownian diffusivity D are also assumed constant. A steady-state component mass balance can be written for component concentration c:

$$u\frac{\partial c}{\partial x}+v\frac{\partial c}{\partial y}=D\frac{\partial^2 c}{\partial y^2}\quad u \text{ and } v \text{ are normal and tangential velocities} \tag{20-56}$$

Further neglecting the first term allows integration from $y = 0$ at the wall (membrane surface) into the boundary layer. At the wall, the net flux is represented by convection into the permeate

$$vc - D\frac{dc}{dy} = vc_p \tag{20-57}$$

This can be further integrated from the wall to the boundary layer thickness $y = \delta$, where the component is at the bulk concentration c_b. Substituting $J = -v$ and $k = D/\delta$, the mass-transfer coefficient yields the stagnant film model [Brian, *Desalination by Reverse Osmosis*, Merten (ed.), M.I.T. Press, Cambridge, Mass., 1966, pp. 161–292]:

$$\frac{v\delta}{D}=\ln\left(\frac{c_b-c_p}{c_w-c_p}\right)=-\frac{J}{k} \tag{20-58}$$

Although allowing for axial variations in v with a constant wall concentration c_w yields Eq. (20-59), as a more rigorous expression applicable to higher concentrations [Trettin et al., *Chem. Eng. Comm.*, **4,** 507 (1980)] the form of Eq. (20-58) is convenient in analyzing a variety of complex behavior.

$$J = 2.046k\left(\frac{c_w-c_p}{c_b-c_p}\right)^{1/3} \tag{20-59}$$

Substituting the intrinsic passage into Eq. (20-58) and rearranging yields an expression for the polarization modulus c_w/c_b:

$$\frac{c_w}{c_b}=\frac{e^{J/k}}{1+S_{\text{int}}\,(e^{J/k}-1)} \tag{20-60}$$

Figure 20-48 shows Wijmans's plot [Wijmans et al., *J. Membr. Sci.*, **109,** 135 (1996)] along with regions where different membrane processes operate (Baker, *Membrane Technology and Applications,* 2d ed., Wiley, 2004, p. 177). For RO and UF applications, $S_{\text{int}} < 1$, and $c_w > c_b$. This may cause precipitation, fouling, or product denaturation. For gas separation and pervaporation, $S_{\text{int}} > 1$ and $c_w < c_b$. MF is not shown since other transport mechanisms besides Brownian diffusion are at work.

Substituting the observed passage into Eq. (20-60) and rearranging yields Eq. (20-61). A plot of the LHS versus J data yields the mass-transfer coefficient from the slope, similar to the Wilson plot for heat transfer-coefficient determination:

$$\ln\left(1-\frac{1}{S_{\text{obs}}}\right)=\ln\left(1-\frac{1}{S_{\text{int}}}\right)-\frac{J}{k} \tag{20-61}$$

A generalized model of transport allowing for component interactions is provided by nonequilibrium thermodynamics where the flux of component i through the membrane J_i [gmol/(cm^2·s)] is written as a first-order perturbation of the chemical potential $d\mu_i/dx$ [cal/(gmol·cm)]:

$$J_i=\sum_i L_i\frac{d\mu_i}{dx} \tag{20-62}$$

and L_i [1/(cal·cm·s)] is a component permeability. The chemical potential gradient provides the driving force for flow and may include contributions from pressure, concentration, electrical field, and temperature. A complementary description of transport is obtained by modeling each transport mechanism in some detail. See each application section for appropriate models.

Application of Eq. (20-62) to the polarization for two components. A and B requires solution of the component balances (Saksena, Ph.D. dissertation, University of Delaware, 1995):

$$u\frac{\partial c_A}{\partial x}+v\frac{\partial c_A}{\partial y}=\frac{\partial}{\partial y}\left(D_A\frac{\partial c_A}{\partial y}+D_{AB}\frac{\partial c_B}{\partial y}\right) \tag{20-63a}$$

$$u\frac{\partial c_B}{\partial x}+v\frac{\partial c_B}{\partial y}=\frac{\partial}{\partial y}\left(D_B\frac{\partial c_B}{\partial y}+D_{AB}\frac{\partial c_A}{\partial y}\right) \tag{20-63b}$$

Larger components with smaller brownian diffusivity polarize more readily. They can exclude smaller components, reducing their concentration at the membrane surface and increasing their retention by the membrane.

Solvent Transport through the Membrane The fluid flux J through a membrane is commonly modeled as an assembly of n_p parallel tubes per unit area with a uniform radius r and length l displaying Hagen-Poiseuille flow:

$$J=\frac{n_p\pi r^4}{8\eta l}\Delta P \tag{20-64}$$

where η is the fluid viscosity and ΔP the pressure drop. The dependence of flow on r^4 means that larger pores take most of the flow.

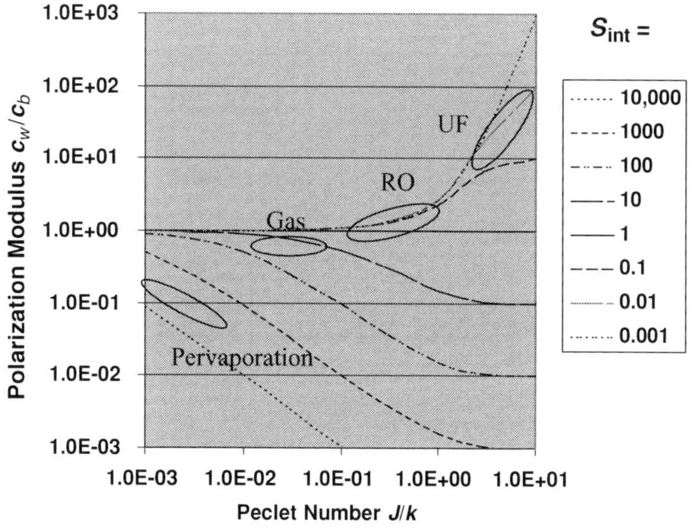

FIG. 20-48 TFF polarization.

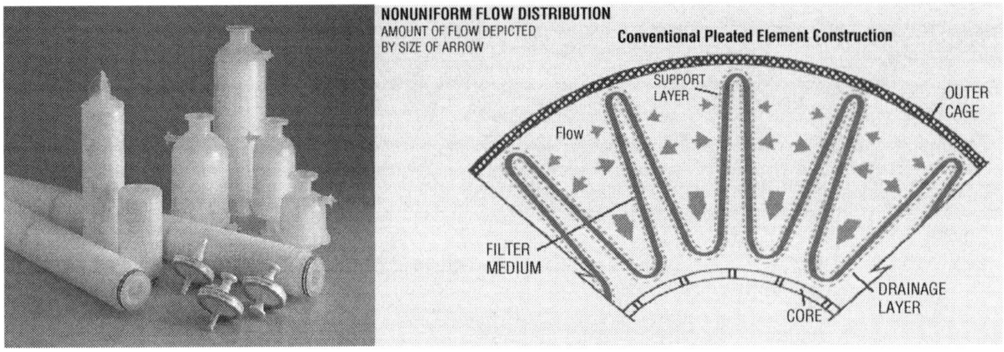

FIG. 20-49 NFF pleated cartridges and capsules. (*Courtesy Millipore Corporation, Pall Corporation.*)

MODULES AND MEMBRANE SYSTEMS

Membranes are packaged into modules for convenient operation. Key module properties include mechanical robustness (providing good sealing between fluid streams and membrane support), good flow distribution (low pressure drops, high TFF membrane shear, low dead volumes, and low sensitivity to feed channel plugging), cleanability, chemical robustness, low extractibles, low cost, ease of assembly, scalability, high product recovery, and high consistency. Cleanability considerations evolved from the dairy industry (3A Sanitary Standards for Crossflow Membrane Modules, USPHS No. 45-00, 1990). Modules may be integrity-tested for the absence of punctures, improper sealing, or other defects.

NFF modules (Figs. 20-49 and 20-50) include cartridges or capsules, stacked disks, and flat sheet. Pleated cartridges, the most common format, are available in a variety of sizes and are installed into multiround stainless-steel housings. As shown in Fig. 20-49, multiple layers of materials such as support layers or membranes with different pore ratings may be pleated together. Capsules are made by sealing individual cartridges within their own plastic housing. Pleated cartridge manufacturing is shown in Fig. 20-51.

TFF module types include plate-and-frame (or cassettes), hollow fibers, tubes, monoliths, spirals, and vortex flow. Figures 20-52 and 20-53 show several common module types and the flow paths within each. Hollow fiber or tubular modules are made by potting the cast membrane fibers or tubes into end caps and enclosing the assembly in a shell. Similar to fibers or tubes, monoliths have their retentive layer coated on the inside of tubular flow channels or lumens with a high-permeability porous structure on the shell side.

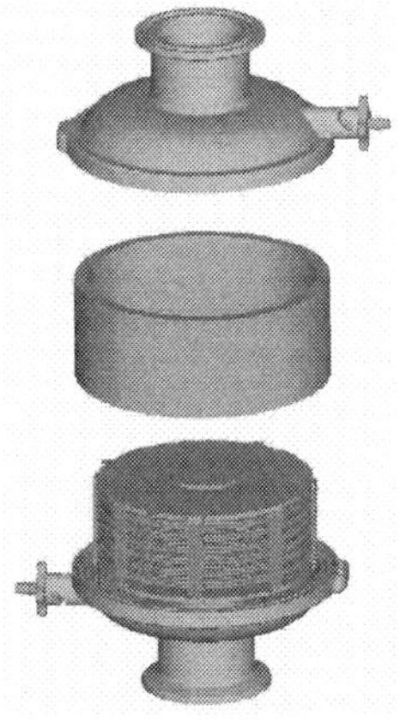

FIG. 20-50 NFF stacked disk modules. (*Courtesy Millipore Corporation.*)

Spiral-wound modules are made by assembling a membrane packet or leaf consisting of a permeate channel spacer sheet with two flat sheets of membranes on either side (facing outward from the permeate sheet). The sides of this packet are glued together, and the top is glued to a central core collection tube containing small holes that allow passage of permeate from the spacer channel into the collection tube. A feed-side spacer is placed on the packet, and the packet is wound around the core to create the module. The wound packet assembly is sometimes inserted into a shell. Spiral modules frequently use a multileaf construction in which several packets are glued to the central core before winding. This design reduces the pressure drop associated with a long permeate path length.

Cassettes use feed and permeate spacers with one set of precut holes for the feed/retentate and another, smaller set of holes for the permeate. The feed spacer has a raised-edge seal around the permeate hole to prevent flow, and the permeate spacer has a similar seal around the feed/retentate hole. Similar to spiral assembly, membrane packets are glued together around their edges and assembled into a stack, and the stack is glued together around the edges.

For fibers or tubular modules, the feed is generally introduced into the inside of the tubes, or lumen, while permeate is withdrawn from the shell side. This flow orientation enhances the shear at the membrane surface for TFF operation. These modules may also be run at high conversion or NFF mode with the feed introduced on the outside of the tubes or shell side. In this case, the shell side offers greater surface area.

Table 20-17 summarizes TFF module characteristics. The preferred TFF module type depends on the characteristics of each application.

In designing TFF modules, the polarization equation [Eq. (20-58)] indicates that for given fixed concentrations, increases in the mass-transfer coefficient k will increase the flux. This can be accomplished by (1) increasing the shear rate at the wall through higher tangential flow velocities and smaller channel heights, or mechanically moving the membrane by spinning or vibration; (2) altering the geometry of the feed channel to increase turbulent mixing in the normal direction by using feed channel screens or curving channels to introduce Dean or Taylor vortices; (3) introducing pulsatile flow in the feed channel or periodic bursts of gas bubbles; and (4) introducing body forces (centrifugal, gravitational, or electromagnetic) to augment transport away from the membrane and introducing large particles in the feed (e.g., >0.5 μm PVC latex beads) that disrupt boundary layers and cause shear-induced diffusion.

Another device that finds frequent use is the stirred cell shown in Fig. 20-54. This device uses a membrane coupon at the bottom of the reservoir with a magnetic stir bar. Stirred cells use low fluid volumes and can be used in screening and R&D studies to evaluate membrane types and membrane properties. The velocity profiles have been well defined (Schlichting, *Boundary Layer Theory*, 6th ed., McGraw-Hill, New York, 1968, pp. 93–99).

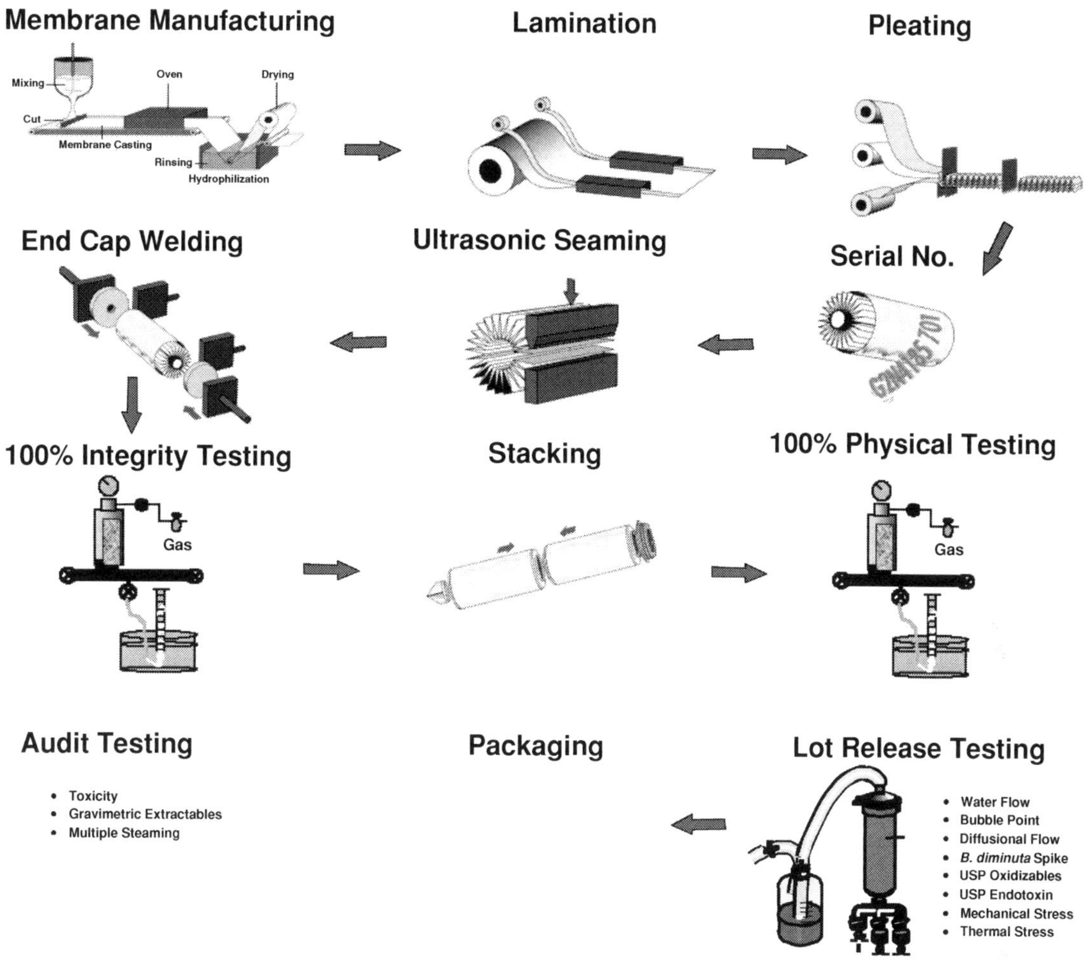

FIG. 20-51 Pleated cartridge manufacturing.

Mass-transfer coefficients for a single newtonian component in various module types and flow regimes can be correlated by Eq. (20-65) with values for the constants in Table 20-18:

$$Sh = a Re^b Sc^c (d_h/L)^e \tag{20-65}$$

where $Sh = kD/d_h$

$$Re = \frac{\rho u d_h}{\mu} \text{ (tube, slit or spacer with hydraulic diameter } d_h\text{)}$$

$$Re = \frac{\rho \omega d^2}{\mu} \text{ (stirred cell diameter } d\text{)}$$

$$Re = \frac{2\rho\omega R d}{\mu} \text{ (rotating-cylinder inner radius } R \text{ and gap diameter } d\text{)}$$

$$Sc = \frac{\mu}{\rho D}$$

(a) (b) (c)

FIG. 20-52 Commercial TFF modules: (*a*) hollow fibers; (*b*) spirals; (*c*) cassettes. (*Courtesy Millipore Corporation.*)

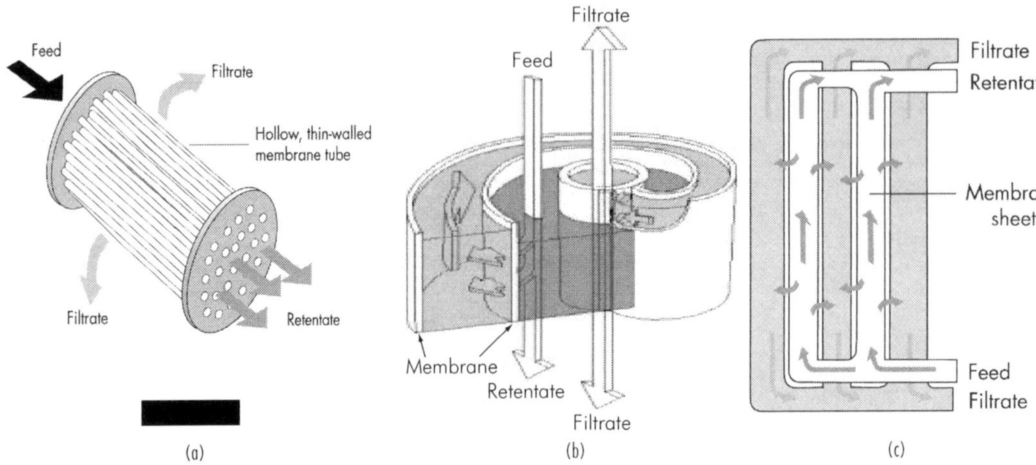

FIG. 20-53 TFF module flow paths: (*a*) hollow fibers; (*b*) spirals; (*c*) cassettes. (*Courtesy Millipore Corporation.*)

Large-Area Configurations NFF membrane systems consist primarily of large, multiround housings that contain pleated cartridges inserted into a common base plate. These cartridges share a common feed and collect a common filtrate.

TFF membrane systems generally use a common feed distributed among parallel modules with a collection of common retentate and common permeate streams. In some applications, it is also useful to plumb TFF modules with the retentate in series where the retentate flow from one module provides the feed flow to the next module. This type of configuration is equivalent to increasing the length of the retentate channel. Permeate flows may or may not be plumbed together.

Process Configurations Basic membrane process configurations shown in Fig. 20-55 include single-pass, batch, fed-batch, and continuous.

Single-Pass Operation A single-pass configuration is used for NFF system operation (no retentate). Component concentrations change along the length of the retentate channel. For dilute streams such as water, assemblies of TFF modules that run in NFF mode (retentate closed) are also used. For nondilute streams, a single pass through a module may be insufficient to generate either the desired permeate flow or a concentrated retentate.

Steady-state component and solvent mass balances can be written for single-pass operation by considering an incremental area element dA in the axial or flow direction for a feed channel, module, or membrane assembly of constant width. For volumetric crossflow Q in the feed channel and observed solute passage S_i,

$$\frac{d(c_iQ)}{dA} = c_i\frac{dQ}{dA} + Q\frac{dc_i}{dA} = -Jc_iS_i \qquad (20\text{-}66)$$

$$\frac{dQ}{dA} = -J \qquad (20\text{-}67)$$

Combining these equations and integrating yield $c_i = c_{io}X^{1-S_i}$ for a volume reduction factor $X = Q/Q_o$ and the observed component passage S_i. This allows one to determine either final concentrations from crossflow rates or the reverse. For a fully retained product ($S_i = 0$), a 10-fold volume reduction ($X = 10$) produces a 10-fold more concentrated product. However, if the product is only partially retained, the volume reduction does not proportionately increase the final concentration due to losses through the membrane.

The area required for processing $A = (Q_o - Q)/\bar{J}$, where $Q_o - Q$ is the permeate volumetric flow, can be estimated by using the approximation $\bar{J} \approx 0.33J_{\text{initial}} + 0.67J_{\text{final}}$ (Cheryan, *Ultrafiltration Handbook*, Technomic, Lancaster, Pa., 1986) and a suitable flux model. An appropriate model relating flux to crossflow, concentration, and pressure is then applied. Pressure profiles along the retentate channel are empirically correlated with flow for spacer-filled channels to obtain $\Delta P = \Delta P_o(Q/A)^n$.

TABLE 20-17 Commercial TFF Modules

	Spiral	Fiber	Cassette
Screens/spacers	Yes	No	Yes/no
Typical no. in series	1–2	1–2	1–2
Packing density, m^2/m^3	800	1000–6000	500
Feed flow, $L/m^2/h$ (LMH)	700–5000	500–18,000	400
Feed pressure drop, psi/module	5–15	1–5	10–50
Channel height, mm	0.3–1	0.2–3	0.3–1
Plugging sensitivity	High	Moderate	High
Working volume, L/m^2	1	0.5	0.4
Holdup volume, L/m^2	0.03	0.03	0.02
Module cost, $\$/m^2$	40–200	200–900	500–1000
Ruggedness	Moderate	Low-moderate	High
Module areas, m^2	0.1–35	0.001–5	0.05–2.5
Membrane types	RO-UF	RO-UF-MF	UF-MF
Relative mass-transfer efficiency*	6	4	10
Ease of use	Moderate	High	Moderate
Scalability†	Fair	Moderate	Good

*Qualitative, based on relative fluxes.
†Cassettes keep retentate path length constant and require lower feed flow rates.

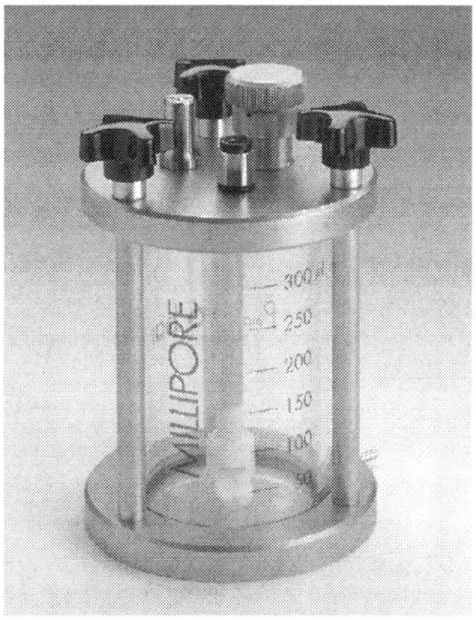

FIG. 20-54 Stirred cell TFF module. (*Courtesy Millipore Corporation.*)

Batch and Fed-Batch Operation Batch operation involves recycling the retentate to the feed tank to create a multipass flow of the feed through the module. The compositions in batch systems change over time. Recycling is necessary for nondilute streams to generate either the desired permeate flow or a concentrated retentate. Multipass operation increases residence times and pump passes that may degrade retentate components. For systems requiring high pressures to generate permeate flow, it is useful to run the entire system at high pressure to save energy.

An unsteady-state component mass balance, Eq. (20-68), can be written for batch operation by assuming a uniform average retentate concentration c_i within the system. Assuming a constant solvent concentration and a 100 percent passage, the solvent balance becomes Eq. (20-69).

$$\frac{d(c_iV)}{dt} = -JAc_iS_i \tag{20-68}$$

$$\frac{dV}{dt} = -JA \tag{20-69}$$

Dividing Eq. (20-68) by Eq. (20-69) and integrating yield the relationship in Table 20-19 between retentate concentration, the volume reduction factor X = (initial retentate volume)/(final retentate volume), and the observed component passage S_i. For a fully retained product ($S_i = 0$), a 10-fold volume reduction ($X = 10$) produces a 10-fold more concentrated product. However, if the product is only partially retained, the volume reduction does not proportionately increase the final concentration due to losses through the membrane. The component mass in the retentate is obtained as the product of the retentate concentration and retentate volume, as shown in Table 20-19.

TFF systems have a maximum volume reduction capability, typically about 40X. This arises because tanks must be large enough to hold the batch volume while allowing operation at a minimum working volume. The minimum working volume may be limited by air entrainment into the feed pump, mixing in the feed tank, or level measurement capabilities. Fed-batch operation extends the volume reduction capability to 100X and provides some flexibility in processing a variety of batch volumes in a single skid. For a fed-batch operation, the retentate is returned to a smaller tank, not the large feed tank. Feed is added to the small retentate tank as permeate is withdrawn, so the system volume remains constant. The smaller retentate tank can allow a smaller working volume without entrainment. While one could make the retentate tank a section of pipe that returns the retentate directly into the pump feed (bypass line), such a configuration can be problematic to flush, vent, and drain, leading to cleaning and product recovery issues.

Equation (20-70) is the unsteady-state component mass balance for fed-batch concentration at constant retentate volume. Integration yields the equations for concentration and yield in Table 20-19.

$$V\frac{dc_i}{dt} = JAc_F - JAc_iS_i \tag{20-70}$$

Retentate concentration over the course of the process is shown in Fig. 20-56 for a fully retentive product ($S_i = 0$). The tank ratio = V_0/retentate tank volume, C_0 is the feed concentration, and V_0 is the feed volume. The benefits sought from higher concentrations can lead to other problems such as reduced fluxes, larger area and pumps, possible denaturation, and extra lines that may have issues with cleaning and product recovery. This not only has caused significant commissioning and validation delays but also has led to the scrapping of a process skid as unworkable. The number of pump passes will also be higher, leading to greater potential protein degradation. Fed batch and bypass should be used only when necessary.

Diafiltration After concentration has removed permeable components from the system, a new buffer can be added to dilute the retained product back to its original volume. Repeated application of this procedure, known as batch diafiltration, is used to exchange one buffer for another and is conveniently implemented at lab scale.

An operating mode known as constant volume diafiltration involves adding a new buffer to the system while withdrawing permeate at the same rate. Figure 20-57 shows the relationship between retentate concentration and diavolumes N = (buffer volumes added)/(fixed retentate volume) for batch and constant volume diafiltration with $S = 1$ (Cheryan, 1998). For a fully passing solute ($S = 1$) such as a buffer, the retentate concentration decays 10-fold with each 2.3 diavolumes. Partially retained solutes do not decay as quickly and require more diavolumes to reach a final concentration target. A fully retained solute ($S = 0$) maintains its retentate concentration constant at the initial value. For fully passing solutes, > 4.6 diavolumes are needed to achieve a specification of < 1 percent of the original buffer

TABLE 20-18 Mass-Transfer Correlations

Geometry	Flow	a	b	c	e	Comments
Tube	Laminar	1.62	0.33	0.33	0.33	Theoretical[1]
Slit	Laminar	1.86	0.33	0.33	0.33	Theoretical[1]
	Turbulent	0.023	0.80	0.33	—	Theoretical[2]
	Turbulent	0.023	0.875	0.25	—	Theoretical[3]
Spacer	—	0.664	0.50	0.33	0.50	Experimental[4]
Stirred cell	Laminar	0.23	0.567	0.33	—	Experimental[5]
Rotating	Laminar	0.75	0.50	0.33	0.42	Experimental[6]

[1]Leveque, *Ann. Mines*, **13,** 201 (1928).
[2]Gekas et al., *J. Membr. Sci.*, **30,** 153 (1987).
[3]Deissler in *Advances in Heat and Mass Transfer*, Harnett (ed.), McGraw-Hill, New York, 1961.
[4]DaCosta et al., *J. Membr. Sci.*, **87,** 79 (1994) with L = spacer mesh length/2, $d_h = 4e/[2/h + (1 - e)s]$ for spacer length channel height h, porosity e, and specific area s. See Costa for additional correction k_{dc}.
[5]Smith et al., *Chem. Eng. Prog. Symp. Ser.*, **64,** 45 (1968).
[6]Holeschovsky et al., *AIChE J.*, **37,** 1219 (1991).

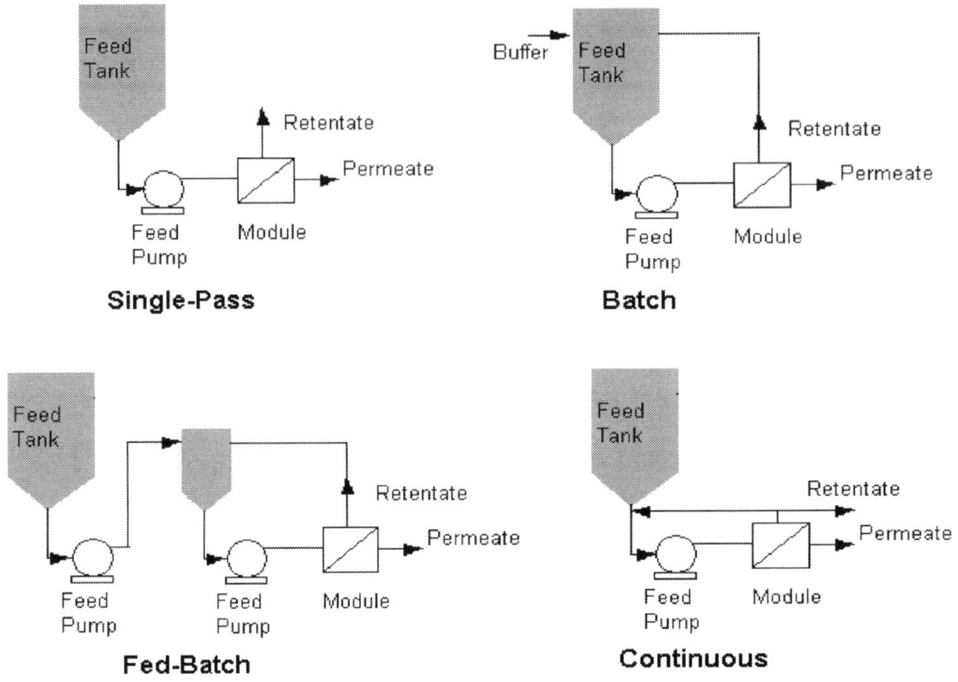

FIG. 20-55 Process configurations.

components. It is common to add an extra 1 to 2 diavolumes as a safety factor to ensure complete buffer exchange. Note that incomplete mixing (due to dead legs and liquid droplets on tank walls) becomes significant at high diavolumes ($N > 10$) and causes the curves in Fig. 20-57 to flatten out.

System sizing involves integration of Eq. (20-69) using a flux model to give Eq. (20-71), where V_p is the permeate volume and J is the average flux. Note the direct tradeoff between area and process time. Table 20-19 shows the concentration and diafiltration steps separated and processing time.

$$A = \frac{V}{\bar{J}t} \quad \text{where} \quad \frac{V_P}{\bar{J}t} = \int_0^{V_P} \frac{dV}{J} \tag{20-71}$$

The formulas for concentration and diafiltration can be combined for the entire process to derive an expression for the loss of product in the permeate. This loss is shown in Fig. 20-58 and Table 20-19 for different levels of processing and membrane retention characteristics ($R = 1 - S$). Note that a membrane with 1 percent passage (99 percent retention) can have yield losses much higher than 1 percent because the protein is repeatedly cycled past the membrane during the entire process with losses at every pass. A high-yielding process (< 1 percent loss) requires membrane passage of < 0.1 percent (retention of > 99.9 percent).

The purification of a product p from an impurity i by an ultrafiltration process is shown in Fig. 20-59 and Eq. (20-72), where C_{i0} and C_{p0} are the initial concentrations (g/L) of the two components, Y_p is the yield of product in the retentate, and ψ = selectivity = S_i/S_p, the ratio of passages. High yields are obtained in purifying out small solutes (high selectivity) but are compromised in removing larger impurities with similar passages to the product.

$$\text{Purification factor PF} = \frac{c_i/c_{i0}}{c_p/c_{p0}} = \left(\frac{\text{ppm in product}}{\text{ppm in feed}}\right) = Y_p^{\Psi-1} \tag{20-72}$$

TABLE 20-19 Batch and Fed-Batch Performance Equations

	Batch concentration and diafiltration	Fed-batch concentration
Retentate concentration	$c_i = c_{i0} X^{1-S_i} e^{-S_i N}$	$c_i = \frac{c_{i0}}{S_i}[1 - (1 - S_i)e^{-S_i(X-1)}]$
Retentate mass	$M_i = M_{i0}\, e^{-S_i(N + \ln X)}$	$M_i = \frac{M_{i0}}{S_i}[1 - (1 - S_i)e^{-S_i(X-1)}]$
Sizing	$A = \frac{V_0}{t}\left(\frac{1 - 1/X}{\bar{J}_{\text{conc}}} + \frac{N/X}{\bar{J}_{\text{diaf}}}\right)$	$A = \frac{V_0}{t}\left(\frac{1 - 1/X}{\bar{J}_{\text{conc}}}\right)$
Comments	• Simplest system and control scheme • Smaller area or process time • Limited to $X < 40$	• Required for $40 < X < 100$ • Minimize fed-batch ratio (= total V_0 / recycle tank V) to minimize area • Can require larger area and time • Integrate for J or use Cheryan approximation • Test process performance using proposed fed-batch ratio

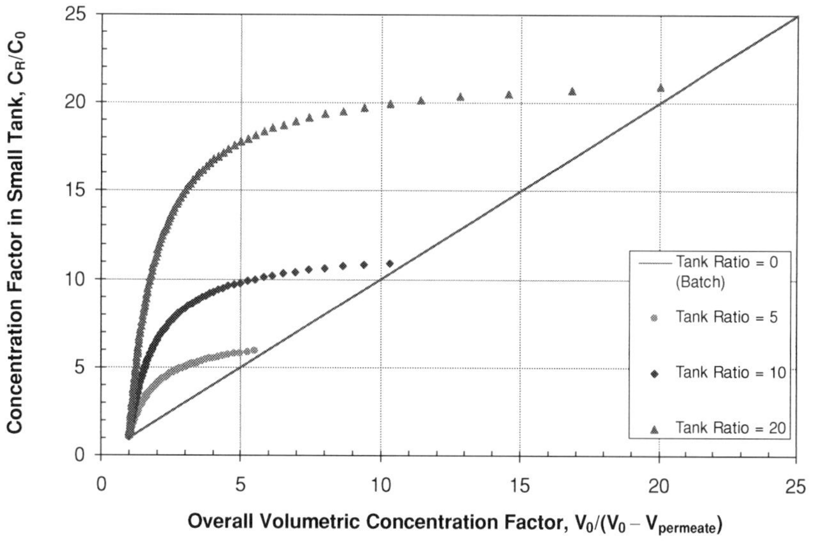

FIG. 20-56 Batch and fed-batch concentration.

Continuous Operation Continuous operation (also called feed-and-bleed) involves the partial recycle of the retentate. The residence time and number of pump passes are in between those of single-pass and batch operation, depending on the fraction of retentate recycled. Several continuous units can be plumbed with the retentate flow feeding the next consecutive system. This configuration is commonly used for large-scale membrane systems.

System Operation NFF system requirements depend on each application but should generally allow for installation, flushing, integrity testing, cleaning or sanitization, processing, recovery, and change-out.

Continued decline in performance indicates a membrane cleaning or compatibility issue. The adequacy of the cleaning step is determined by the recovery of at least 80 percent of the initial normalized water flux. Although some variability in water flux is typical, any consistent decline reflects an inadequate cleaning procedure.

Process Development/Scale-up Feasibility is evaluated by using small single-module systems. This involves selection of membrane type, operating processing conditions, processing strategies (concentration, diafiltration, etc.), cleaning procedures, performance evaluation, and sensitivity to feedstock variations. Pilot operation employs a larger skid, representative of planned commercial operation, running representative feedstock over extended periods to assess longer-term repeatable performance (product quality and retention, costs, yields, process flux, and membrane integrity).

Module change-out is based on preset number of uses, time, or when performance (retention, flux, integrity) drops below preset specifications.

Fluxes scale directly and volumes scale proportionately with the feed volume. A safety factor will be built into the scale-up design so that process times will not be the same.

REVERSE OSMOSIS (RO) AND NANOFILTRATION (NF)

RO and NF processes employ pressure driving forces of 0.3 to 10.5 MPa to drive liquid solvents (primarily water) through membranes while retaining small solutes. Reverse-osmosis (RO) membranes are commonly rated by a > 90 percent NaCl retention and retention of > 50 Da neutral organics while passing smaller liquid solvents. Nanofiltration membranes (sometimes called low-pressure RO, loose RO, or hyperfiltration) have 20 to 80 percent NaCl retention and retain > 200 to 1000 Da neutral organics. Dissolved gases have low retention. Neutral or undissociated solutes have lower retention than charged or dissociated solutes. Although virus-retaining membranes are sometimes referred to as nanofilters, in keeping with membrane scientist classification they are described under ultrafiltration. RO membranes will remove bacteria and viruses. Lab-scale sample preparation in dilute solutions can be run as NFF in centrifuge tubes. TFF is used in large commercial processes and small home and shipboard water purification systems.

Commercial interest in RO began with the first high-flux, high-NaCl-retention Loeb-Sourirajan anisotropic cellulose acetate membrane. Practical application began with the *thin film composite* (*TFC*) membrane and implementation for seawater desalination at Jeddah, Saudi Arabia [Muhurji et al., *Desalination*, **76,** 75 (1975)].

Applications RO is primarily used for water purification: seawater desalination (35,000 to 50,000 mg/L salt, 5.6 to 10.5 MPa operation), brackish water treatment (5000 to 10,000 mg/L, 1.4 to 4.2 MPa operation), and *low-pressure RO* (*LPRO*) (500 mg/L, 0.3 to 1.4 MPa operation). A list of U.S. plants can be found at www2.hawaii.edu, and a 26 Ggal/yr desalination plant is under construction in Ashkelon, Israel. Purified water product is recovered as permeate while the concentrated retentate is discarded as waste. Drinking water specifications of total dissolved solids (TDS) ≤ 500 mg/L are published by the U.S. EPA and of ≤ 1500 mg/L by the WHO [Williams et al., chap. 24 in *Membrane Handbook*, Ho and Sirkar (eds.), Van Nostrand, New York, 1992]. Application of RO to drinking water is summarized in Eisenberg and Middlebrooks (*Reverse Osmosis Treatment of Drinking Water*, Butterworth, Boston, 1986).

RO competes well with multistage flash evaporation in desalination applications and is the current benchmark due to advantages in both capital and operating costs. It may be too costly to compete against freshwater sourced from wells and pipelines. LPRO and electrodialysis compete for brackish water treatment applications. Compact home kitchen RO runs on 0.4-MPa line pressure for < 2000-mg/L TDS feed solutions to reduce salts, odor, and trace metals such as As. These units range from \$200 to \$500 (U.S. 2005) to produce water at \$0.06 to \$0.25 per gallon. RO processes were estimated to represent 23 percent of the 3 Ggal/day of worldwide potable water production in 1986 [Wangnick, *IDA Magazine*, **1,** 53 (1986)].

Useful conversions are as follows:

Volume: 1 m^3 = 1000 L = 264.2 U.S. gal = 8.11×10^{-4} acre·ft = 35.32 ft^3
Pressure: 1 bar = 14.5 psi = 0.1 MPa
Flow: 1 U.S. gal/min = 1440 gal/day = 227 L/h = 5.47 m^3/day
Concentration: 1 mg/L = 1 ppm = 17.1 grains/gal

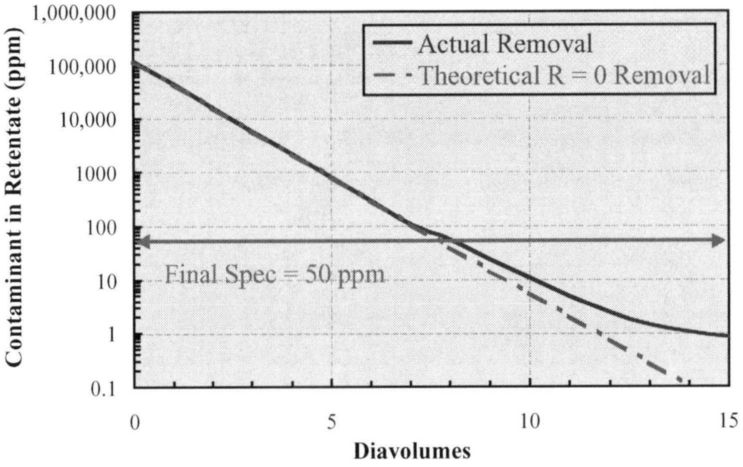

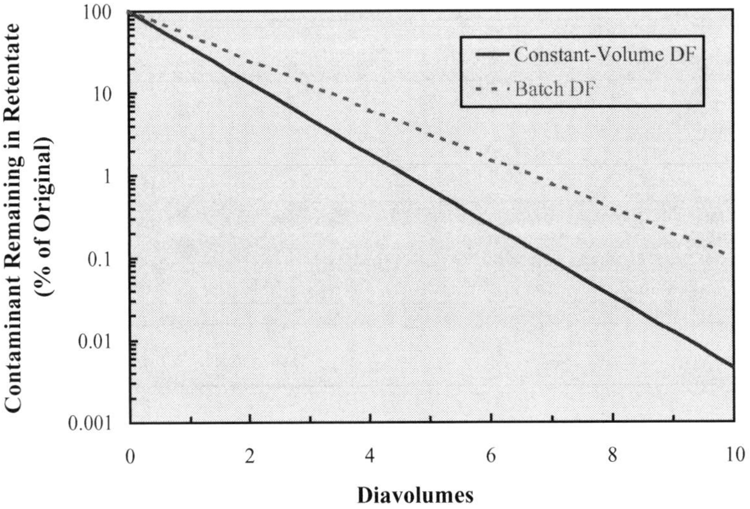

FIG. 20-57 Diafiltration performance (Cheryan, 1998).

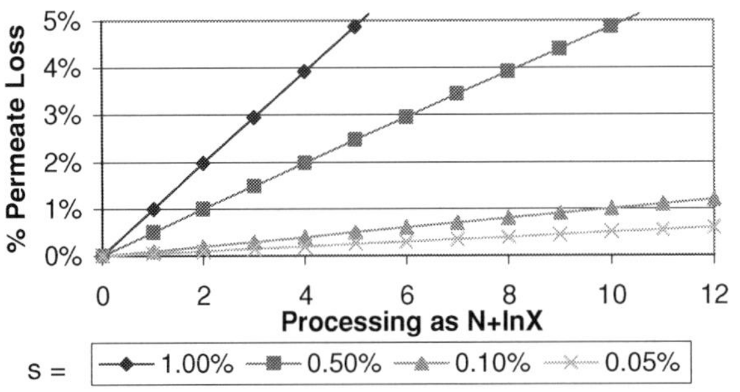

FIG. 20-58 Permeate losses.

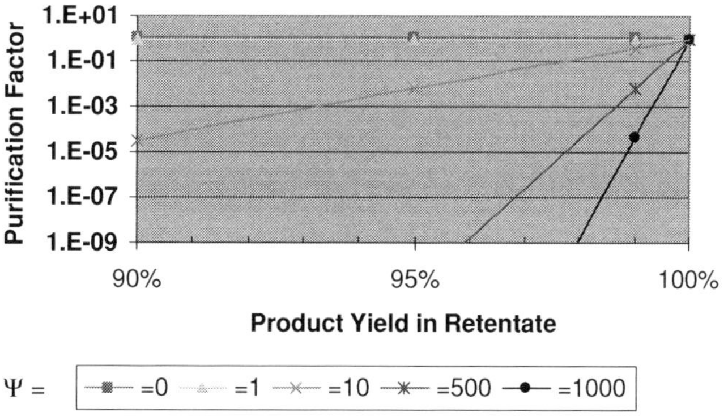

FIG. 20-59 Purification versus yield.

Process water applications include boiler water feed pretreatment before ion exchange or electrodialysis. RO is also used for ultrahigh-purity water production for use in laboratories, medical devices (kidney dialysis), microelectronic manufacturing (rinse fluids per ASTM D-19 D5127-90, 1990), and pharmaceutical manufacturing (purified water or water for injection as specified by USP).

RO applications recovering valuable product in the retentate include tomato, citrus, and apple juice dewatering, sometimes in combination with thermal evaporators at \$0.15 to \$0.18 per gallon [Short, chap. 9 in *Bioseparation Processes in Foods*, Singh and Rezvi (eds.), Marcel Dekker, New York, 1995]. Dealcoholization of wine or beer involves diafiltering alcohol through RO membranes while retaining flavor compounds in the retentate (Meier, *Wine East*, Nov./Dec., 10, 1991). LPRO is used to desalt soy sauce and dairy whey. Small-molecule drugs (e.g., penicillin) can be recovered by RO membranes.

RO can recover metals, antifreeze, paint, dyes, and oils in the retentate while generating cleaned up wastewater permeate for disposal. RO is also used to reduce the volume of waste liquids (e.g., spent sulfite liquor in paper manufacturing). Wastewater treatment application removals of 95 percent TOC, > 90 percent COD, > 98 percent PAH compounds, and pesticides > 99 percent have been seen [Williams et al., chap. 24 in *Membrane Handbook*, Sirkar and Ho (eds.), Van Nostrand, 1992].

Membranes, Modules, and Systems

Membranes RO membranes are designed for high salt retention, high permeability, mechanical robustness (to allow module fabrication and withstand operating conditions), chemical robustness (to fabrication materials, process fluids, cleaners, and sanitizers), low extractables, low fouling characteristics, high capacity, low cost, and consistency.

The predominant RO membranes used in water applications include cellulose polymers, *thin film composites (TFCs)* consisting of aromatic polyamides, and crosslinked polyetherurea. Cellulosic membranes are formed by immersion casting of 30 to 40 percent polymer lacquers on a web immersed in water. These lacquers include cellulose acetate, triacetate, and acetate-butyrate. TFCs are formed by interfacial polymerization that involves coating a microporous membrane substrate with an aqueous prepolymer solution and immersing in a water-immiscible solvent containing a reactant [Petersen, *J. Membr. Sci.*, **83,** 81 (1993)]. The Dow FilmTec FT-30 membrane developed by Cadotte uses 1-3 diaminobenzene prepolymer crosslinked with 1-3 and 1-4 benzenedicarboxylic acid chlorides. These membranes have NaCl retention and water permeability claims.

Cellulose membranes can generally tolerate a pH of 3 to 6 and 0.3 to 1.0 ppm of chlorine while TFC membranes can generally tolerate a pH of 3 to 11 and < 0.05 ppm of chlorine. Membrane tolerance is also described as the permitted cumulative ppm-hours of membrane exposure to chlorine. TFC membranes can range from 1000 to 12,000 ppm·h. Always check specific filter specifications. RO membranes approved for food use by the FDA include aromatic polyamides, polypiperazinamide, and aryl-alkyl polyetherurea (Code of Federal Regulations, Title 21, Part 177, U.S. GPO, Washington, D.C., annually).

Nanofiltration membranes are negatively charged and reject multivalent anions at a much higher level than monovalent anions, an effect described as Donnan exclusion. Nanofiltration membranes have $MgSO_4$ retention and water permeability claims.

Modules RO modules are available in spiral, hollow fiber, tubular, and plate and frame formats. A comparison of characteristics is shown in Table 20-20. Spirals are used the most since they are compact, easy-to-use modules; have low feed-side pressure drops; are less prone to clogging; and are easily cleaned, mechanically robust, and economical. The standard spiral module has an 8-in diameter, is 40 in long with about 30-m^3 area. An anti-telescoping end cap is used to prevent the leaves from being pushed out of alignment by axial retentate flow. Hollow fibers are run with the feed on the shell side, flowing radially inward and out along the lumen. This places less mechanical stress on the fibers. Tubular modules are used with high-suspended-solids and high-viscosity fluids.

System Configurations Figure 20-60 shows common RO system configurations for permeate product recovery. For single-pass operation, the feed passes once through a membrane to generate acceptable permeate product. Double-pass operation uses permeate staging (the permeate from the first module feeds the second module) with allowance for blending to generate acceptable product. This configuration may be needed for high-concentration, hard-to-remove impurities, high product purities, and lower-retention modules. Double-pass operation has been recommended for desalination of > 38,000 mg/L [Klinko et al., *Desalination*, **54,** 3 (1985)] or modules with salt retention < 99.2 percent [Kaschemekat et al., *Desalination*, **46,** 151 (1983)]. It is also common in ultrahigh-purity water systems.

Spirals are normally packed at 5 to 7 modules per cylindrical housing and operate with staged retentates where the retentate from the first module feeds the second module. Retentate staging increases the length of the feed channel. Retentate staging of housings is used to increase the conversion ratio from around 25 to 50 percent in one housing stage to 50 to 75 percent for two housing stages and 75 to 90 percent in three housing stages. Since the feed flow is reduced by the

TABLE 20-20 RO Module Performance

	Spiral	Hollow fiber	Tubular	Plate and frame
Packing density, m^2/m^3	800	6000	70	500
Feed flow, m^3/(m^2·s)	0.3–0.5	0.005	1–5	0.3–0.5
Feed ΔP, psi	40–90	1.5–4.5	30–45	45–90
Plugging sensitivity	High	High	Low	Moderate
Cleanability	Fair	Poor	Excellent	Good
Relative cost	Low	Low	High	High
Chemistries available	C and TFC	C and TFC	C only	C and TFC

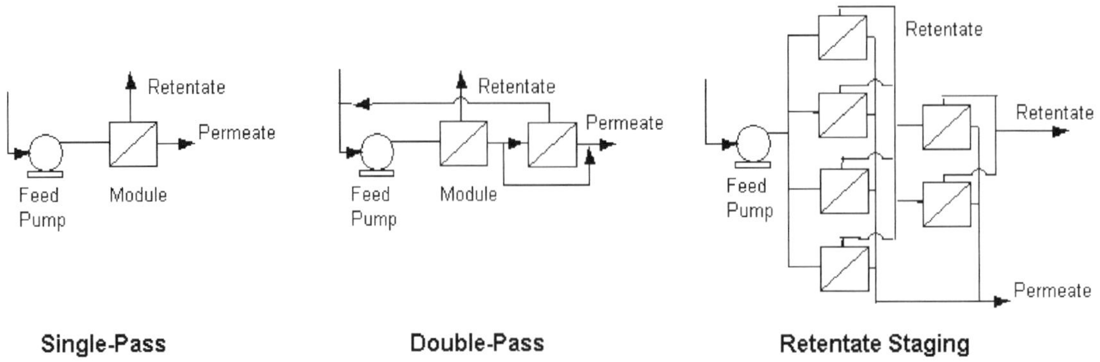

FIG. 20-60 RO system configurations.

conversion ratio after passing through the first module stage, maintaining a minimum crossflow rate requires that the second module stage have less membrane area in what has been called a "Christmas tree" design. Interstage pumping between housings may be required to maintain pressures and flows.

Retentate product recovery employs the batch, fed-batch, or continuous processing systems, and concentration or diafiltration modes of operation (see general membrane section). Beer and wine dealcoholization uses batch diafiltration.

Component Transport in Membranes RO is analyzed as a rate, not equilibrium, process. Polarization of retained salts and other low-MW components is described in the general membrane section. Solvent flux J_w and solute flux J_i through RO membranes can be described by using Eq. (20-73), the solution diffusion imperfection model [Sherwood et al., *Ind. Eng. Chem. Fund.*, **6,** 2 (1967)]. This model considers transport through the membrane in parallel with transport through a small number of nonretentive defects. Transport of solvent or solute through the membrane occurs by partitioning between the upstream polarization layer and membrane phases at the membrane surface, diffusion through the membrane, and partitioning between the membrane and downstream fluid. This process is reflected in permeabilities P_w and P_2. Dissolution in the membrane and diffusion through it can be modeled as an activated process with an associated inverse exponential temperature dependence of the permeabilities. Transport of solvent or solute through defects occurs by convective flow. Water viscosity increases roughly 2 percent/°C near 25°C.

$$J_w = \frac{P_w}{l}(\Delta P - \Delta\pi) + \frac{P_3}{l}\Delta P \qquad J_i = \frac{P_2}{l}(c_w - c_p) + \frac{P_3}{l}\Delta P c_w \qquad (20\text{-}73)$$

In accordance with observed data, this model shows that water flux J_w increases linearly with applied pressure ΔP, decreases with higher salt concentration through its impact on osmotic pressure π, increases with a smaller membrane thickness l, and increases with temperature through the temperature dependence of the water permeability P_w. The model also demonstrates that the solute or salt flux J_i increases linearly with applied pressure ΔP, increases with higher salt concentration c_w, increases with a smaller membrane thickness l, and increases with temperature through the temperature dependence of the solute permeability P_2. Polarization, as described early in this section, causes the wall concentration c_w to exceed the bulk concentration c_b.

Osmotic Pressure The osmotic pressure π of salt solutions is calculated from

$$\pi = -\frac{RT}{\overline{V}_w}\ln a_w = \frac{RTM_w}{1000\overline{V}_w}\sum_i m_i\phi_i \qquad (20\text{-}74)$$

for water activity a_w, water partial molal volume $\overline{V}_w$, gas constant R, absolute temperature T, water molecular weight M_w, ion molality m_i, and osmotic coefficient ϕ_i for each ionic species i. Note that both Na^+ and Cl^- are counted as separate species for NaCl. For seawater at 35,000 to 50,000 mg/L (ppm) and 25°C, the osmotic pressure is in the range of 2.7 to 3.8 MPa (~11 kPa/1000 ppm).

Solute Retention Retention is determined by the relative rates of solute and solvent transport through the membrane. The impact of operation on solute retention $R_i = 1 - c_p/c_w$ can be evaluated from Eq. (20-73) by using the mass balance $J_i = c_p J_w$. Assuming high retention, $c_p << c_w$, and the solvent flux through defects is negligible, so the retention is calculated as

$$R_i = 1 - \frac{J_i}{J_w c_w} = 1 - \frac{P_2 + P_3\Delta P}{P_w(\Delta P - \Delta\pi)} \qquad (20\text{-}75)$$

At low pressures where $\Delta P \sim \Delta\pi$, solvent flux is low but solute flux is high (it is insensitive to pressure and occurs under a concentration gradient). This results in low solute retention. Pressures of 1 MPa over the feed osmotic pressure are needed to maintain high retention. Increasing the applied pressure ΔP increases the solvent flux, diluting the solute flux with solvent and resulting in a higher retention. Increasing the feed solute concentration lowers the solvent flux through its impact on osmotic pressure π, and decreases retention. Increasing temperature increases the solvent permeability P_w, the solute permeability P_2, and the osmotic pressure. Solute retention tends to drop slightly at higher temperature.

Ionic solutions require charge neutrality to prevent a higher flux of species of one charge causing an associated electric current and voltage. This may limit the net salt flux through the membrane by the ion with the lowest permeability so that the sodium ion will show different retentions depending on the anion it is paired with. Charge neutrality can also enrich the permeate concentration of a salt relative to the feed to create a negative retention. For example, a feed solution containing Na^+, Li^+, Cl^-, and SO_4^{2-} ions would show a permeate with enriched levels of Cl^- to balance permeated Na^+ and Li^+ ions.

Nanofilters incorporate negative membrane charge for higher anion rejection. High feed salinities can passivate these charges and reduce anion retention.

Pretreatment and Cleaning Pretreatment is commonly used to extend membrane life and increase recovery. The representative pretreatment train for water purification applications in Fig. 20-61 controls feed channel clogging, mineral scaling, fouling by organic films and microorganisms, and oxidants that can degrade the membranes.

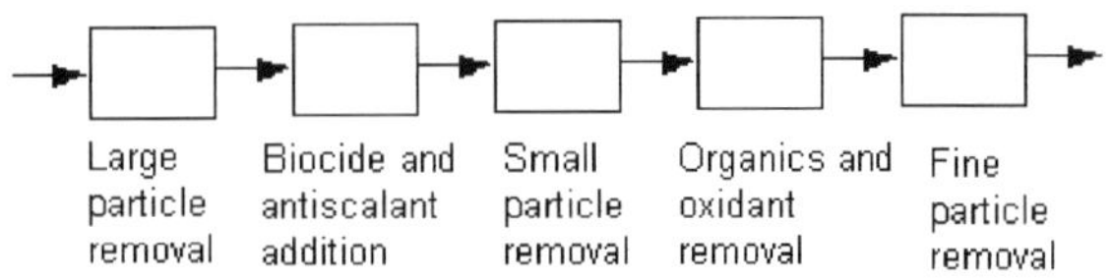

FIG. 20-61 RO pretreatment schematic.

TABLE 20-21 SDI Interpretation

SDI value	Operability
<1	Can run several years without colloidal fouling
1–3	Can run several months between cleaning
3–5	Frequent cleaning required
>5	Pretreatment required to reduce colloids

Particle Removal Solid particles can plug feed channel spacer channels and deposit as a cake layer on the membrane surface, causing increased flow resistance and lowering fluxes. Table 20-19 indicates the sensitivity of different module types to feed channel plugging. Particles may consist of organic colloids, rust flakes, precipitated iron hydroxide, algae, clay, colloidal silica, or silt. Pretreatment includes using a combination of coarse screening, hydrocyclones, flocculent addition (e.g., alum, ferric sulfate, polyelectrolytes), a sand or multimedia filtration bed with periodic backflushing, and depth filter cartridges or ultrafiltration systems.

The ability of a feed solution to plug a membrane is determined by the *silt density index* (*SDI*) employing a 0.45-μm PVDF microfilter at 30-psi differential pressure (ASTM Standard D-4189-82, 1987).

$$\text{SDI} = \frac{100(1 - T_i/T_f)}{15} \qquad (20\text{-}76)$$

where T_i is the time it takes to filter the first 0.5 L and T_f is the time it takes to filter 0.5 L starting when 15 min of test time has elapsed. Table 20-21 is an interpretation of SDI values.

Antiscalants Precipitation of salts can form a scale on the membrane surface when solubility limits (Table 20-22) are exceeded in the retentate. This decreases flux.

Metal hydroxides (e.g., Fe, Mn, Al) can also be a problem (Rautenbach and Albrecht, *Membrane Processes*, Wiley, New York, 1989). A chemical analysis of the feed solution composition along with consideration of solubility products allows one to determine the significance of precipitation. Solubility products can be affected by temperature, pH, and ionic strength. Seasonal temperature variations must be considered. Concentrations of silica need to be < 120 mg/L in the feed.

Water hardness in milligrams per liter of total Ca^{2+} and Mg^{2+} is considered soft at 0 to 17 mg/L, moderately hard at 60 to 120 mg/L, and very hard at > 180 mg/L. The Langelier saturation index, a measure of $CaCO_3$ solubility, should be < 0 in the retentate to prevent precipitation (*Pure Water Handbook*, Osmonics, 1997).

Scale prevention methods include operating at low conversion and chemical pretreatment. Acid injection to convert CO_3^{2-} to CO_2 is commonly used, but cellulosic membranes require operation at pH 4 to 7 to prevent hydrolysis. Sulfuric acid is commonly used at a dosing of 0.24 mg/L while hydrochloric acid is to be avoided to minimize corrosion. Acid addition will precipitate aluminum hydroxide. Water softening upstream of the RO by using lime and sodium zeolites will precipitate calcium and magnesium hydroxides and entrap some silica. Antiscalant compounds such as sodium hexametaphosphate, EDTA, and polymers are also commonly added to encapsulate potential precipitants. Oxidant addition precipitates metal oxides for particle removal (converting soluble ferrous Fe^{2+} ions to insoluble ferric Fe^{3+} ions).

Oxidant Removal The presence of oxidizers such as chlorine or ozone can degrade polyamide RO membranes, causing a drop in salt retention. Cellulosic membranes are less sensitive to attack. Addition of 1.5 to 6 mg sodium bisulfite/ppm chlorine or contacting with activated carbon will remove oxidizers. Vacuum degassing with a hydrophobic filter module is also used.

Biocides Microorganisms can form biofilms on membrane surfaces, causing lower fluxes [Flemming in *Reverse Osmosis*, Z. Amjad (ed.), Van Nostrand, New York, 1993, p. 163]. Cellulosic membranes can be consumed by bacteria, causing lower retention. Control involves pretreatment with chemicals or UV light and membrane chemical sanitization on a regular or periodic basis. Chemical sanitants include chlorine, bromine, ozone and peroxide/peracetic acid oxidants (subject to membrane limits), formaldehyde, sodium bisulfite, copper sulfate for algae and plankton control, or chloramines. The use of oxidizers will also remove dissolved hydrogen sulfide. Cellulosic membranes are commonly sanitized with 0.2 mg/L of free chlorine.

TABLE 20-22 Solubility Limits at 30°C

$CaCO_3$	$MgSO_4$	$CaSO_4$	$BaSO_4$	SiO_2	$SrSO_4$
65 mg/L	9 mg/L	2090 mg/L	2 mg/L	120 mg/L	114 mg/L

Organic Foulants Organic components can range from 1 mg/L in the open ocean to 80 to 100 mg/L in coastal waters. Components include humic and fulvic acids (800 to 50,000 kDa) from decayed vegetation, or contaminating petroleum that can adsorb to the membranes and reduce their flux. Carbon beds used for dechlorination can also remove these organics by adsorption. For wastewater purification to remove organics, high pH and ozonation improve retention and reduce flux decline (Bhattacharyya et al., *Separation of Hazardous Organics by LPRO—Phase II*, U.S. EPA, 1990).

Cleaning/Sanitization Membrane cleaning restores initial membrane flux without compromising retention to extend membrane life. Appropriate treatment depends on the nature of the foulants, membrane compatibility, and compatibility of the chemicals with the application. Cleaning generally uses acid flushing (for scalants) but may include base or enzyme flushing (for organics) and 80°C hot water flushing (for sanitization/microorganism control).

Design Considerations and Economics

Osmotic Pinch As the feed travels along the feed channels, friction losses lower the feed pressure. Solvent flux along the feed channel dewaters the feed and increases solute concentration and osmotic pressure. For a recovery or conversion ratio CR, a module solute retention R, and an inlet solute concentration of c_o, the bulk concentration c_b is

$$c_b = c_o(1 - \text{CR})^{-R} \qquad (20\text{-}77)$$

Both the feed pressure drop and the osmotic pressure increase lower the driving force for solvent flow $\Delta P - \Delta\pi$ along the feed channel, and the point at the end of the feed channel, having the lowest driving force, is the *osmotic pinch point*. This point has both the lowest solvent flux and the lowest solute retention. If the driving force falls below 0 along the feed channel, the solvent flux will reverse (from permeate into the feed) and subsequent membrane area will perform "negative work." Economic justification of membrane at the pinch point requires that it generate adequate flux to pay for itself. This, in turn, requires that the driving force be about 1 MPa.

Process Economics and Optimization Table 20-23 shows some representative process economics for several large 1 Mgal/day system designs [adapted from Ray, chap. 25 in *Membrane Handbook*, Ho and Sirkar (eds.),Van Nostrand, New York, 1992]. These costs are a useful guide, within ±25 percent, but specific quotes are needed to account for site-specific differences between projects. No energy recovery technologies are included. For capacities above 50 to 100 gal/day, system costs scale almost linearly, as (gal/day capacity)$^{0.85\text{–}0.95}$, and can be conveniently expressed as dollars per gallon per day. Total costs are split evenly between capital and operating costs.

Energy use (pumping high flows to high pressures) represents the biggest operating cost while membrane replacement operating costs are only 10 to 15 percent. These operating costs can be represented as

$$\$E\ (\$/1000\text{ gal}) = (0.724\text{ kWh/psig-1000 gal}) \times (\Delta P\text{ psid})(\$/\text{kWh})/\text{CR-Ef} \qquad (20\text{-}78)$$

where CR is the fractional conversion ratio (recovery), Ef is the pump efficiency, and

$$\$A\ (\$/1000\text{ gal}) = (0.432)(\$/\text{m}^2)/\text{CR-}J\text{-}T \qquad (20\text{-}79)$$

where J is the average system flux, L/(m^2·h), and T is the years of membrane life.

By incorporating a flux model [Eq. (20-73)], solute concentration [Eq. (20-77)], and polarization model [Eq. (20-60)] into the area cost

TABLE 20-23 Representative RO Process Costs

Costs	Seawater	Brackish water	Softening	Notes
Operating conditions				
Inlet pressure	6.9 MPa	3.1 MPa	1.4 MPa	
Flux	25 LMH*	42 LMH	42 LMH	
Conversion	40%	60%	75%	
Total cost, \$/1000 gal	4.7	2.3	1.3	15-yr life
Capital cost	2.1	1.1	0.6	12% interest
Operating cost	2.6	1.2	0.7	15% downtime
Total capital cost, \$/(gal/day)	4.5	2.3	1.3	
Direct costs	3.7	1.9	1.1	
Equipment	3.3	1.6	0.8	
Indirect costs	0.8	0.4	0.2	
Total operating cost, \$/1000 gal	2.6	1.2	0.7	
Energy	1.6	0.4	0.2	\$0.075/kWh
Membrane replacement	0.4	0.2	0.2	3-yr life
Chemicals	0.2	0.2	0.1	2 per shift @\$14/h
Labor	0.3	0.3	0.2	
Other	0.1	0.1	0.0	

*LMH = liters per square meter per hour.

[Eq. (20-79)], the operating cost term ($\$A + \$E$) varies with ΔP and CR. The osmotic pinch constraint requires a minimum ΔP to overcome the average osmotic pressure and requires a maximum CR where the concentrated retentate osmotic pressure exceeds the feed pressure. Assuming that solute retention is sufficient to maintain permeate quality, there exist optimum conversion ratios and ΔP that depend on the feed (osmotic pressure), module (mass transfer, permeability, life, cost), and pump (energy cost, efficiency). Conversion ratios per module stage range from 25 to 50 percent, and typical inlet pressures are shown in Table 20-23. No provision has been made here for costs of disposing of the retentate.

One can also rearrange Eq. (20-79) to solve for ΔP in terms of area and substitute in Eq. (20-78) to express the operating costs in terms of membrane area. The energy operating cost is inverse in area, while the membrane replacement cost is linear, reflecting the existence of an optimum area. Addition of capital costs leads to total costs that scale as $b(\text{area}^{0.95}) + c(\text{area}^{-0.6})$. Module vendors provide good general recommendations and support to design and economic evaluation.

Recognizing that the high-pressure retentate contains considerable kinetic energy before it is passed through a retentate valve provides an opportunity for energy recycling. A variety of energy recovery technologies have been proposed including Pelton wheels, reverse-running pumps, hydraulic turbochargers, and pressure exchangers with pumping energy reductions of 22 to 60 percent [Darwish et al., *Desalination*, **75**, 143 (1989)]. Energy recovery technologies are most useful for high-pressure desalination but may not be worth the investment for brackish water and LPRO applications (Glueckstern et al., chap. 12 in *Synthetic Membrane Processes*, Academic Press, New York, 1984).

The capital cost of a 15 kgal/day RO skid, uninstalled with instrumentation, is about \$35,000 (U.S., 2005) and scales with gallons-per-day capacity to the 0.52 power [Ray, chap. 25 in *Membrane Handbook*, Sirkar and Ho (eds.), Van Nostrand, New York, 1992]. Approximately 25 to 30 percent of this cost is associated with membranes and housings. Operating costs are roughly \$0.30/1000 gal for energy and membrane replacement, and \$0.09/1000 gal for pretreatment, maintenance, and cleaning. High-purity water systems employing double-pass staging operate at conversions of 70 to 80 percent in the first pass and 70 to 75 percent in the second pass. Home units using line pressure run at 5 to 15 percent conversion.

System Operation For corrosive pressurized seawater, 316L stainless steel is commonly used for piping and housings although specialized alloys may be needed (Oldfield et al., *Desalination*, **55**, 261 (1985)]. Sensors are used to measure inlet pH and temperature (to control scaling), pump pressure (to prevent surges and ensure sufficient driving force), pump flow (to ensure crossflow to prevent excessive polarization), permeate flow (to monitor performance), and permeate conductivity (to ensure module integrity and product quality). Control systems are used to monitor and display sensor values, track and optimize operation over time, trigger alarms, and store the data. Provisions for module replacement and start-up are needed.

For batch processing, systems comparable to ultrafiltration are used.

Posttreatment of the permeate for potable water use can include dissolved CO_2 removal to prevent corrosion (by aeration, lime treatment), chlorination for microbial control, and oxygenation to improve taste.

ULTRAFILTRATION

Ultrafiltration processes (commonly UF or UF/DF) employ pressure driving forces of 0.2 to 1.0 MPa to drive liquid solvents (primarily water) and small solutes through membranes while retaining solutes of 10 to 1000 Å diameter (roughly 300 to 1000 kDa). Commercial operation is almost exclusively run as TFF with water treatment applications run as NFF. Virus-retaining filters are on the most open end of UF and can be run as NFF or TFF. Small-scale sample preparation in dilute solutions can be run as NFF in centrifuge tubes.

The first large commercial application of UF was paint recycling, followed by dairy whey recovery in the mid-1970s. UF applications are enabled by low-temperature and low-cost operation.

Applications Electrodeposition of cationic paint resin on automobiles (connected to the cathode) provides a uniform, defect-free coating with high corrosion resistance, but carries with it about 50 percent excess paint that must be washed off. UF is used to maintain the paint concentration in the paint bath while generating a permeate that is used for washing. The spent wash is fed back into the paint path (Zeman et al., *Microfiltration and Ultrafiltration*, Marcel Dekker, New York, 1996).

During cheese making, the coagulated milk or curd is used to make cheese while the supernatant whey is a waste product rich in salts, proteins, and lactose. Whey concentration and desalting by UF produce a retentate product that can be used as an animal feed supplement or food additive. The MMV process (Maubois et al., French Patent 2,052,121) involves concentrating the milk by UF after centrifugation to remove the cream and before coagulation to improve yields and reduce disposal costs.

UF is used in biopharmaceutical purification (proteins, viral and bacterial vaccines, nucleic acids) for initial concentration of clarified fluids or lysates (to reduce subsequent column sizes and increasing binding), to change buffer conditions (for loading on to columns), and to concentrate and change buffers (for final formulation into a storage buffer) [Lutz et al., in *Process Scale Bioseparations for the Biopharmaceutical Industry*, Shukla, et al. (eds.), CRC Press, New York, 2006]. For these applications, the drug product is recovered in the retentate. For drug products with potential viral contaminants

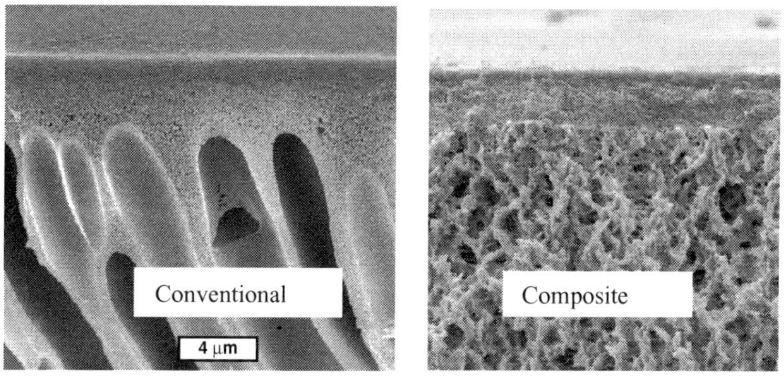

FIG. 20-62 Ultrafiltration membranes. (*Courtesy Millipore Corporation.*)

(sourced from mammalian cells or human and animal blood plasma), virus removal filters employed in either a TFF or NFF mode retain the viruses while the drug product is recovered in the permeate.

UF is used to clarify various fruit juices (apple, grape, pear, pineapple, cranberry, orange, lemon) which are recovered as the permeate [Blanck et al., *AIChE Symp. Ser.* **82,** 59 (1986)]. UF has also been used to remove pigments and reducing browning in wine production [Kosikowski in *Membrane Separations in Biotechnology*, McGregor (ed.), Marcel Dekker, New York, 1986].

Wastewater treatment and water purification applications employ UF in a TFF or NFF mode to produce permeate product with reduced colloids, pyrogens, and viruses. Oil droplets in wastewater are retained by UF for recycle or disposal at a significantly reduced volume.

Membranes, Modules, and Systems

Membranes UF membranes consist primarily of polymeric structures (polyethersulfone, regenerated cellulose, polysulfone, polyamide, polyacrylonitrile, or various fluoropolymers) formed by immersion casting on a web or as a composite on a MF membrane. Hydrophobic polymers are surface-modified to render them hydrophilic and thereby reduce fouling, reduce product losses, and increase flux [Cabasso in *Ultrafiltration Membranes and Applications*, Cooper (ed.), Plenum Press, New York, 1980]. Some inorganic UF membranes (alumina, glass, zirconia) are available but only find use in corrosive applications due to their high cost.

Early ultrafiltration membranes had thin surface retentive layers with an open structure underneath, as shown in Fig. 20-62. These membranes were prone to defects and showed poor retention and consistency. In part, retention by these membranes would rely on large retained components in the feed that polarize or form a cake layer that plugs defects. Composite membranes have a thin retentive layer cast on top of a microfiltration membrane in one piece. These composites demonstrate consistently high retention and can be integrity-tested by using air diffusion in water.

Table 20-24 compares properties of commonly used polyethersulfone (PES) or regenerated cellulose membranes. Membrane selection is based on experience with vendors, molecular weight rating for high yields, chemical and mechanical robustness during product processing and clean-in-place (extractables, adsorption, swelling, shedding, class VI), process flux for sizing and costing, and the quality/consistency (ISO, cGMP) of the vendor and the membrane. Membranes are described by their normal water permeability (NWP), the ratio of their flux at 25°C to the average transmembrane pressure (TMP). Measurements taken at different temperatures can be adjusted to 25°C using a viscosity ratio correction. Process fluxes will be somewhat lower due to fouling. Membranes containing a hydrophilic surface generally have lower fouling and produce higher process fluxes.

UF membranes are assigned a nominal molecular weight limit (NMWL) corresponding roughly to their ability to retain a solute of a particular molecular weight. While tight membranes (i.e., low-molecular-weight cutoff ratings) provide high retention, they have corresponding low flow rates requiring greater membrane area to do the job. This leads to larger pumps with large holdup volumes and potential negative impact on product quality. A rule of thumb for selecting membrane NMWL is to take 0.2 to 0.3 of the MW of the solute that is to be retained by the membrane. However, the NMWL values should only be used as a very rough guide or screening method to select membranes for testing. Retention is based on hydrodynamic size, not molecular weight, so linear chain dextrans show a higher passage than globular proteins of the same molecular weight. In addition, there is no standardization of rating methods so specific retention properties vary considerably among membranes and vendors depending on the marker solute selected (e.g., protein, dextran in a particular buffer) and the level of retention selected for the marker solute (90, 95, 99 percent). Retention is affected by fouling due to adsorbed components and polarized solutes on the membrane surface. See retention below for additional discussion.

Modules and Systems UF modules include cassettes, spirals, hollow fibers, tubes, flat sheets, and inorganic monoliths. These are primarily run in TFF operation to increase flux by reducing plugging. Exceptions run in NFF operation include virus removal and water treatment applications where both cartridges and hollow fibers are used. Continuous operation provides lower-cost operation, but the biologicals process requires batch processing to meet regulatory requirements. Applications with high solids require multiple-pass operation to obtain a significant conversion ratio, but water treatment

TABLE 20-24 Ultrafiltration Membrane Properties

Material	Advantages	Disadvantages
Polyethersulfone (PES)	Resistance to temperature, Cl_2, pH, easy fabrication	Hydrophobic
Regenerated cellulose	Hydrophilic, low-fouling	Sensitive to temperature, pH, Cl_2, microbial attack, mechanical creep
Polyamide		Sensitive to Cl_2, microbial attack
Inorganic	Resistance to temperature, Cl_2, pH, high pressure, long life	Cost, brittleness, high crossflow rates

TABLE 20-25 UF Modules and Systems

Application	Paint recovery	Whey processing	Biologicals	Water treatment	Juice processing
Membrane	PES	PES	Cellulose, PES, PVDF	PES	PES, PVDF
Module	Fibers	Spirals	Cassette	Fibers	Tubes
System	Continuous	Continuous	Batch, NFF	Continuous, single-pass	Batch
Characteristics	Cost, plugging	Cost	Recovery	Cost	Plugging, cost

can achieve high conversion in one pass. The properties of these modules and system operation are described in the general section. Table 20-25 indicates where these modules are used.

Component Transport in Membranes UF is analyzed as a rate, not equilibrium, process. Pure solvent transport in UF is characterized by $J = \text{TMP}/R_{\text{membrane}}$, where TMP (transmembrane pressure) is the pressure difference across the membrane and R is the membrane resistance. Resistance R can be related to membrane properties by using Eq. (20-64) with pore size distributions following a lognormal distribution [Zydney et al., *J. Membr. Sci.*, **91**, 293 (1994)]. At ionic strengths < 0.1 M, charged membranes generate a streaming potential across their thickness which can lead to a 15 percent or more diminished solvent flux by counterelectroosmosis (Newman, *Electrochemical Systems*, Prentice-Hall, Englewood Cliffs, N.J., 1973). Most UF process fluids are aqueous with higher ionic strengths, but permeability measurements using low-conductivity water can encounter this effect.

Figure 20-63 shows that in the presence of retained solutes, the solvent flux flattens out at high pressures. UF has a high level of polarization where the membrane surface concentration can range up to 100 times the bulk concentration. Retained antibodies at a wall concentration of 191 g/L have an osmotic pressure Π of 30 psi (Mitra, 1978). That is, an elevated pressure of 30 psig must be applied to the protein-rich retentate side of a water-permeable membrane containing 191 g/L of antibody in order to prevent water backflow from the permeate side of the membrane containing water at 0 psig. This diminishes the driving force for flow and leads to the mechanistic-based osmotic flux model (Vilker, 1984):

$$\text{Polarization flux model} \quad J = \frac{\text{TMP} - R \cdot \Delta\Pi(c_w)}{\mu \cdot (R_{\text{membrane}} + R_{\text{fouling}})} \tag{20-80}$$

where R is the intrinsic membrane retention and $\Delta\Pi(C_w)$ is the osmotic pressure at the wall concentration. An empirically based flux model can also be defined by omitting the osmotic term and adding a compressible polarization resistance term in the denominator (Cheryan, 1998).

Equation (20-80) requires a mass transfer coefficient k to calculate c_w and a relation between protein concentration and osmotic pressure. Pure water flux obtained from a plot of flux versus pressure is used to calculate membrane resistance (typically small). The LMH/psi slope is referred to as the NWP (normal water permeability). The membrane plus fouling resistances are determined after removing the reversible polarization layer through a buffer flush. To illustrate the components of the osmotic flux model, Fig. 20-63 shows flux versus TMP curves corresponding to just the membrane in buffer ($R_{\text{fouling}} = 0$, $c_w = 0$), fouled membrane in buffer ($c_w = 0$), and fouled membrane with osmotic pressure.

The region at low flux/low TMP is called the linear region and is dependent on TMP but independent of crossflow and bulk concentration. The region at high flux/high TMP is called the polarized region and is independent of TMP but dependent on crossflow and bulk concentration. This extremely counterintuitive result is the consequence of polarization. In between these two regions lies what is termed the "knee" of the flux curve. Increasing pressure beyond the knee gives diminishing returns in improving flux.

In the polarized region, the empirical gel model of Eq. (20-81) can provide a good description of the flux behavior. The gel model is based on polarization [Eq. (20-58)] with a constant wall concentration C_g and complete retention. The mass-transfer coefficient k is dependent on crossflow velocity, as shown in Table 20-18 for different geometries and flow regimes. Concentration-dependent transport (viscosity and diffusivity) will alter the velocity exponents. Operating at the knee of the flux curve, either by design or due to fouling that causes a shift in the flux curve to the right, will decrease the velocity dependence. Note that the empirical gel point concentration does not correspond to an actual gel on the membrane surface. As the feed concentration rises to within 5 to 10 percent of the gel point, the flux tends to drop off less rapidly than Eq. (20-81) would predict. The gel point is insensitive to crossflow but can increase with temperature. Cheryan summarizes gel point and velocity exponent values for a variety of solutions.

$$J = k \ln c_g/c \tag{20-81}$$

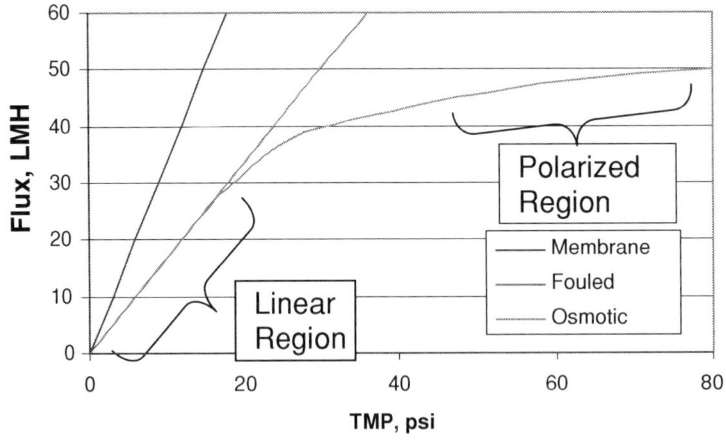

FIG. 20-63 Ultrafiltration flux behavior.

Solute Flux Solute partitioning between the upstream polarization layer and the solvent-filled membrane pores can be modeled by considering a spherical solute and a cylindrical pore. The equilibrium partition coefficient ϕ (pore/bulk concentration ratio) for steric exclusion (no long-range ionic or other interactions) can be written as

Solute partitioning $$\phi = (1 - \lambda)^2 \quad (20\text{-}82)$$

where λ = solute radius/pore radius.

Solute flux within a pore can be modeled as the sum of hindered convection and hindered diffusion [Deen, *AIChE J.*, **33,** 1409 (1987)]. Diffusive transport is seen in dialysis and system start-up but is negligible for commercially practical operation. The steady-state solute convective flux in the pore is $J_s = K_cJc = \phi K_cJc_w$, where c is the radially averaged solute concentration and K_c is the convective hindrance factor (from pore wall drag, solute lag, and long-range solute pore wall and solute-solute potential interactions). For uncharged cylindrical pores $\phi K_c = \phi(2 - \phi)\exp(-0.7146\lambda^2)$ to within 2 percent [Zeman et al., in *Synthetic Membranes,* vol. 2, Turbak (ed.), *ACS Symp. Ser.,* **54,** ACS, Washington, D.C. (1981), p. 412].

A charged solute in a pore has a higher potential energy due to the distortion of its electrical field when it penetrates the pore wall into the polymer. This effect increases solute retention. The apparent size of solutes can be magnified by more than an order of magnitude by this effect. See Deen for other pore geometries and the impacts of long-range solute pore wall and solute-solute interactions. The use of electrostatic interactions to separate solutes is discussed by van Reis [*Biotechn. and Bioeng.*, **56**(1), 71 (1997)].

Solute Passage/Retention The intrinsic passage (c_p/c_w) is determined by the ratio of the solute flux to the solvent flux as $s_i = \phi K_c$. The intrinsic passage is inherent to the membrane and solute. As shown in Fig. 20-64 for a single pore, solute passage (or retention) does not follow a step change. Figure 20-65 shows passage for a membrane with a pore size distribution [Tkacik et al., *Biotechnology*, **9,** 941 (1991)].

The observed passage (c_p/c_b) varies with both the intrinsic passage and the extent of polarization, as shown in Eq. (20-83):

Observed passage $$s_o = \frac{1}{1 + \left(1/s_i - 1\right)\cdot \exp(-J/k)} \quad (20\text{-}83)$$

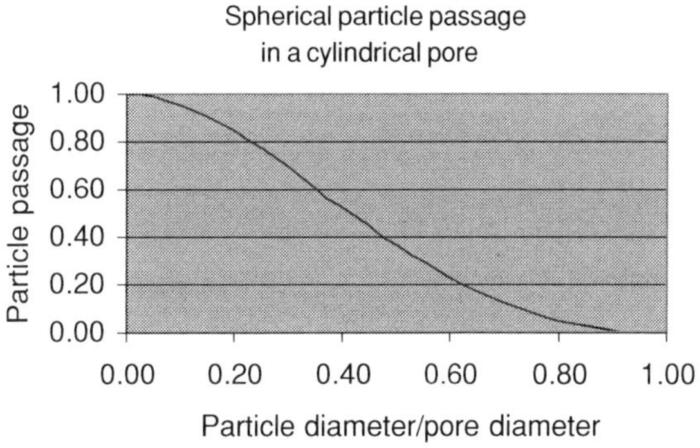

FIG. 20-64 Solute passage.

At low flux these passages are equivalent. At high fluxes the wall concentration is high due to polarization and the observed passage increases, approaching 100 percent, regardless of the intrinsic passage.

In multicomponent systems, large solutes with lower diffusivity, polarize more and exclude smaller solutes from the membrane surface, decreasing their passage. Operation at the "knee" of the flex curve reduces this effect.

Design Considerations and Economics The selection among batch, fed-batch, single-pass, or continuous operation has been described earlier in this section. Important performance considerations for UF applications involve product purity and/or concentration, product yield, and cost/sizing for a given production rate.

Batch or Fed-Batch Operation This mode of operation is typical of biologicals and juice processing where high solids and low fluxes require multiple passes, and batch operation is characteristic of the manufacturing process. Formulas in Table 20-19 can be used to calculate the required volume reduction factor X, diafiltration volumes

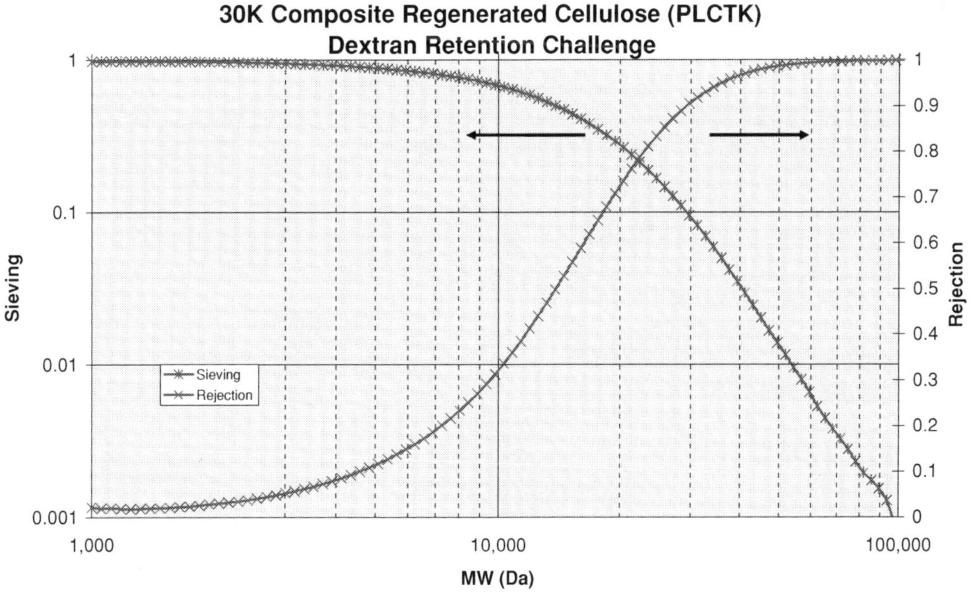

FIG. 20-65 Intrinsic membrane retention.

N, and solute passage S_i needed to produce desired retentate product with impurity concentrations c_i and retentate product yield M_i/M_{i0}. Permeate product characteristics for batch operation can be determined by mass balances using a permeate volume of $V_p = V_0(1 - 1/X + N/X)$, a mass of solute i in the permeate as $M_{i,\text{permeate}} = M_{i0}(1 - e^{-S_i(N+\ln X)})$ and the permeate concentration as the ratio of the mass to volume. Fed-batch equations are obtained by using the same approach.

The system area A needed for batch or fed-batch operation can be calculated by using the formulas in Table 20-19 for production rates V_0/t based on feed volume V_0 and average fluxes during process steps. The final retentate batch volume $V_R = V_0/X$ or permeate batch volume $V_p = V_0(1 - 1/X + N/X)$ can be used to restate the production rate on other bases. Although experience can be used to estimate solute passage and process fluxes, they should be determined experimentally for each application.

Single-Pass TFF Single-pass TFF or NFF operation is typical of water treatment. Low solids allow high fluxes with high permeate product recovery and low retentate disposal costs. NFF operation is also used for virus retention in small-batch biologicals processing where low NFF fluxes (high areas) are compensated by ease of use, low residence time, and low capital costs.

Equations (20-66) and (20-67) present single-pass formulas relating retentate solute concentration, retentate crossflow, permeate flow, and membrane area. For relevant low-feed-concentration applications, polarization is minimal and the flux is mainly a function of pressure. Spiral or hollow fiber modules with low feed channel and permeate pressure drops are preferred.

Continuous Operation This mode of operation is typical of paint recovery, whey processing, and wastewater processing where high solids and low fluxes require multiple passes, and continuous operation is allowed by the manufacturing process. For a feed concentration c_F, a volumetric concentration factor X (= retentate flow/feed flow), and retention R, the outlet retentate concentration is

$$c = c_F\left[\frac{X}{X - R(X-1)}\right] \tag{20-84}$$

In comparison to batch processing, this process operates at the final concentration with associated low fluxes. The use of multiple stages where the retentate outlet from one stage feeds the next allows successive systems to operate at different fluxes and will improve membrane productivities.

MICROFILTRATION

Process Description Microfiltration (MF) separates particles from true solutions, be they liquid or gas phase. Alone among the membrane processes, microfiltration may be accomplished without the use of a membrane. The usual materials retained by a microfiltration membrane range in size from several μm down to 0.2 μm. At the low end of this spectrum, very large soluble macromolecules are retained by a microfilter. Bacteria and other microorganisms are a particularly important class of particles retained by MF membranes. Among membrane processes, dead-end filtration is uniquely common to MF, but cross-flow configurations are often used.

Brief Examples Microfiltration is the oldest and largest membrane field. It was important economically when other disciplines were struggling for acceptance, yet because of its incredible diversity and lack of large applications, it is the most difficult to categorize. Nonetheless, it has had greater membrane sales than all other membrane applications combined throughout most of its history. The early success of microfiltration was linked to an ability to separate microorganisms from water, both as a way to detect their presence, and as a means to remove them. Both of these applications remain important.

Laboratory Microfiltration membranes have countless laboratory uses, such as recovering biomass, measuring particulates in water, clarifying and sterilizing protein solutions, and so on. There are countless examples for both general chemistry and biology, especially for analytical procedures. Most of these applications are run in dead-end flow, with the membrane replacing a more conventional medium such as filter paper.

Medical MF membranes provide a convenient, reliable means to sterilize fluids without heat. Membranes are used to filter injectable fluids during manufacture. Sometimes they are inserted into the tube leading to a patient's vein.

Process Membrane microfiltration competes with conventional filtration, particularly with diatomaceous earth filtration in general-process applications. A significant advantage for membrane MF is the absence of a diatomaceous earth residue for disposal. Membranes have captured most of the final filtration of wine (displacing asbestos), are gaining market share in the filtration of gelatin and corn syrup (displacing diatomaceous earth), are employed for some of the cold pasteurization of beer, and have begun to be used in the pasteurization of milk. Wine and beer filtration operate dead-end; gelatin, corn syrup, and milk are cross-flow operations. MF is used to filter all fluid reactants in the manufacture of microcircuits to ensure the absence of particulates, with point-of-use filters particularly common.

Gas Phase Microfiltration plays an important and unique role in filtering gases and vapors. One important example is maintaining sterility in tank vents, where incoming air passes through a microfilter tight enough to retain any microorganisms, spores, or viruses. A related application is the containment of biological activity in purge gases from fermentation. An unrelated application is the filtration of gases, even highly reactive ones, in microelectronics fabrication to prevent particulates from contaminating a chip.

Downstream Processing Microfiltration plays a significant role in downstream processing of fermentation products in the pharmaceutical and bioprocessing industry. Examples are clarification of fermentation broths, sterile filtration, cell recycle in continuous fermentation, harvesting mammalian cells, cell washing, mycelia recovery, lysate recovery, enzyme purification, vaccines, and so forth.

MF Membranes Microfiltration is a mature field that has proliferated and subdivided. The scope and variety of MF membranes far exceeds that in any other field. A good overview is given by Strathmann [in Porter (ed.), op. cit., pp. 1–78]. MF membranes may be classified into those with tortuous pores or those with capillary pores. Tortuous-pore membranes are far more common, and are spongelike structures. The pore openings in MF are much larger than those in any other membrane. Surface pores may be observed by electron microscope, but tortuous pores are much more difficult to observe directly. Membranes may be tested by bubble-point techniques. Many materials not yet useful for tighter membranes are made into excellent MF membranes. Retention is the primary attribute of an MF membrane, but important as well are permeability, chemical and temperature resistance, dirt capacity (for dead-end filters), FDA-USP approval, inherent strength, adsorption properties, wetting behavior, and service life.

Membrane-production techniques listed below are applicable primarily or only to MF membranes. In addition, the Loeb-Sourirjain process, used extensively for reverse osmosis and ultrafiltration membranes, is used for some MF membranes.

Membranes from Solids Membranes may be made from microparticles by sintering or agglomeration. The pores are formed from the interstices between the solid particles. The simplest of this class of membrane is formed by sintering metal, metal oxide, graphite, ceramic, or polymer. Silver, tungsten, stainless steel, glass, several ceramics, and other materials are made into commercial membranes. Sintered metal may be coated by TiO_2 or zirconium oxide to produce MF and UF membranes. Membranes may be made by the careful winding of microfibers or wires.

Ceramic Ceramic membranes are made generally by the sol-gel process, the successive deposition of ever smaller ceramic precursor spheres, followed by firing to form multitube monoliths. The diameter of the individual channels is commonly about 2 to 6 mm. Monoliths come in a variety of shapes and sizes. A 19-channel design is common. One manufacturer makes large monoliths with square channels.

Track-Etched Track-etched membranes (Fig. 20-66) are now made by exposing a thin polymer film to a collimated beam of radiation strong enough to break chemical bonds in the polymer chains. The film is then etched in a bath which selectively attacks the damaged polymer. The technique produces a film with photogenic pores,

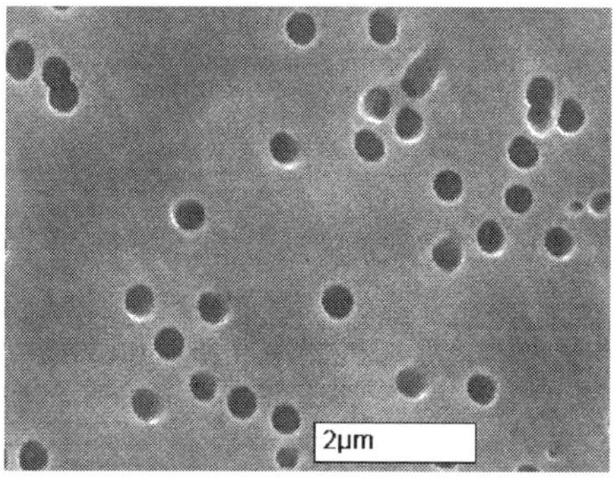

FIG. 20-66 Track-etched 0.4-μm polycarbonate membrane. (*Courtesy Millipore Corporation.*)

whose diameter may be varied by the intensity of the etching step. Commercially available membranes have a narrow pore size distribution and are reportedly resistant to plugging. The membranes have low flux, because it is impossible to achieve high pore density without sacrificing uniformity of diameter.

Chemical Phase Inversion Symmetrical phase-inversion membranes (Fig. 20-67) remain the most important commercial MF membranes produced. The process produces tortuous-flow membranes. It involves preparing a concentrated solution of a polymer in a solvent. The solution is spread into a thin film, then precipitated through the slow addition of a nonsolvent, usually water, sometimes from the vapor phase. The technique is impressively versatile, capable of producing fairly uniform membranes whose pore size may be varied within broad limits.

Thermal Phase Inversion Thermal phase inversion is a technique which may be used to produce large quantities of MF membrane economically. A solution of polymer in poor solvent is prepared at an elevated temperature. After being formed into its final shape, a sudden drop in solution temperature causes the polymer to precipitate. The solvent is then washed out. Membranes may be spun at high rates using this technique.

Stretched Polymers MF membranes may be made by stretching (Fig. 20-68). Semicrystalline polymers, if stretched perpendicular to the axis of crystallite orientation, may fracture in such a way as to make reproducible microchannels. Best known are Goretex® produced from Teflon®, and Celgard® produced from polyolefin. Stretched polymers have unusually large fractions of open space, giving them very high fluxes in the microfiltration of gases, for example. Most such materials are very hydrophobic.

Membrane Characterization MF membranes are rated by flux and pore size. Microfiltration membranes are uniquely testable by direct examination, but since the number of pores that may be observed directly by microscope is so small, microscopic pore size determination is mainly useful for membrane research and verification of other pore-size-determining methods. Furthermore, the most critical dimension may not be observable from the surface. Few MF membranes have neat, cylindrical pores. Indirect means of measurement are generally superior. Accurate characterization of MF membranes is a continuing research topic for which interested parties should consult the current literature.

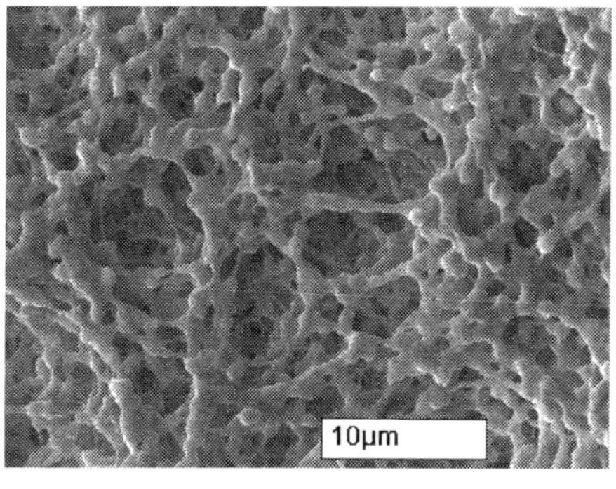

FIG. 20-67 Chemical phase inversion 0.45-μm polyvinylidene fluoride membrane. (*Courtesy Millipore Corporation.*)

FIG. 20-68 Stretched polytetrafluoroethylene membrane. (*Courtesy Millipore Corporation.*)

Bubble Point Large areas of microfiltration membrane can be tested and verified by a bubble test. Pores of the membrane are filled with liquid, then a gas is forced against the face of the membrane. The Young-Laplace equation, $\Delta P = (4\gamma \cos \Theta)/d$, relates the pressure required to force a bubble through a pore to its radius, and the interfacial surface tension between the penetrating gas and the liquid in the membrane pore. γ is the surface tension (N/m), d is the pore diameter (m), and P is transmembrane pressure (Pa). Θ is the liquid-solid contact angle. For a fluid wetting the membrane perfectly, $\cos \Theta = 1$.

By raising the gas pressure on a wet membrane until the first bubble appears, the largest pore may be identified, and its size computed. This is a good test to run on a membrane apparatus used to sterilize a fluid, since bacteria larger than the identified largest pore (or leak) cannot readily penetrate the assembly. Pore-size distribution may also be run by bubble point. Bubble-point testing is particularly useful in assembled microfilters, since the membrane and all seals may be verified. Periodic testing ensures that the assembly retains its integrity. Diffusional flow of gas is a complication in large MF assemblies. It results from gas dissolving in pore liquid at the high-pressure side, and desorbing at the low-pressure side. If the number of pores and the average pore length are known, the effect can be computed. Special protocols are used when this method is used for critical applications. Detail is provided in ASTM F316-86, "Standard test method for pore size characteristics of membrane filters by bubble point and mean flow pore test." The bubble-point test may also be run using two liquids. Because interfacial surface tensions of liquids can be quite low, this technique permits measurements on pores as small as 10 nm.

Charged Membranes The use of tortuous-flow membranes containing a positive electrical charge may reduce the quantity of negatively charged particles passing even when the pore size is much larger than the particle. The technique is useful for making prefilters or layered membranes that withstand much higher solids loadings before becoming plugged.

Bacteria Challenge Membranes are further tested by challenge with microorganisms of known size: their ability to retain all of the organisms is taken as proof that all pores are smaller than the organism. The best-known microorganism for pore-size determination is *Pseudomonas diminuta,* an asporogenous gram-negative rod with a mean diameter of 0.3 μm. Membranes with pore size smaller than that are used to ensure sterility in many applications. Leahy and

Sullivan [*Pharmaceutical Technology,* **2**(11), 65 (1978)] provide details of this validation procedure.

Membrane thickness is a factor in microbial retention. Tortuous-pore membranes rated at 0.22 μm typically have surface openings as large as 1 μm (Fig. 20-67). Narrower restrictions are found beneath the surface. In challenge tests, *P. diminuta* organisms are found well beneath the surface of a 0.2-μm membrane, but not in the permeate.

Latex Latex particles of known size are available as standards. They are useful to challenge MF membranes.

Process Configuration As befits a field with a vast number of important applications and a history of innovation, there are countless variations on how an MF process is run.

Dead-end versus Cross-flow Conventional filtration is usually run dead-end, and is facilitated by amendments that capture the particulates being removed. Membranes have very low dirt capacity, so only applications with very low solids to be removed are run in conventional dead-end flow. A rough upper limit to solids content is about 0.5 percent; streams containing <0.1 percent are almost always processed by dead-end devices. Since dead-end membrane equipment is much less expensive than cross-flow, great ingenuity is applied to protecting the critical membrane pores by structured prefilters to remove larger particles and debris. The feed may also be pretreated. It is common practice to dispose of the spent membrane rather than clean it. The membrane may be run inverted. A review of dead-end membrane filtration is given by Davis and Grant [in Ho and Sirkar (eds.), op. cit., pp. 461–479].

Cross-flow is the usual case where cake compressibility is a problem. Cross-flow microfiltration is much the same as cross-flow ultrafiltration in principle. In practice, the devices are often different. As with UF, spiral-wound membranes provide the most economical configuration for many large-scale installations. However, capillary devices and cassettes are widely employed, especially at smaller scale. A detailed description of cross-flow microfiltration had been given by Murkes and Carlsson [*Crossflow Filtration,* Wiley, New York (1988)].

Membrane Inverted Most membranes have larger openings on one face than on the other. Common practice is to run the tightest face against the feed in order to avoid plugging of the backing by particles. The rationale is that anything that makes it past the "skin" will have relatively unimpeded passage into the backing and out with the permeate. For very low solids this convention is reversed, the rationale being that the porous backing provides a trap for particulates, rather like filter aid.

If the complete passage of soluble macromolecules is required, a highly polarized membrane is an advantage. The upside-down membrane hinders the back diffusion of macrosolutes. Countering the tendency of the retained particulates to "autofilter" soluble macrosolutes, the inhibition of back diffusion raises the polarization and thus the passage of macrosolutes such as proteins. The particulates are physically retained by the membrane. Blinding and plugging can be controlled if the membrane is backwashed frequently. This technique has been demonstrated at high solids loadings in an application where high passage of soluble material is critical, the microfiltration of beer [Wenten, Rasmussen, and Jonsson, *North Am. Membrane Soc. Sixth Annual Meeting,* Breckenridge, CO (1994)].

Liquid Backpulse Solid membranes are backwashed by forcing permeate backward through the membrane. Frequent pulsing seems to be the key.

Air Backflush A configuration unique to microfiltration feeds the process stream on the shell side of a capillary module with the permeate exiting the tube side. The device is run as an intermittent dead-end filter. Every few minutes, the permeate side is pressurized with air. First displacing the liquid permeate, a blast of air pushed backward through the membrane pushes off the layer of accumulated solids. The membrane skin contacts the process stream, and while being backwashed, the air simultaneously expands the capillary and membrane pores slightly. This momentary expansion facilitates the removal of imbedded particles.

Process Limitations The same sorts of process limitations affecting UF apply to MF. The following section will concentrate on the differences.

Concentration Polarization The equations governing cross-flow mass transfer are developed in the section describing ultrafiltration. The velocity, viscosity, density, and channel-height values are all similar to UF, but the diffusivity of large particles (MF) is orders-of-magnitude lower than the diffusivity of macromolecules (UF). It is thus quite surprising to find the fluxes of cross-flow MF processes to be similar to, and often higher than, UF fluxes. Two primary theories for the enhanced diffusion of particles in a shear field, the inertial-lift theory and the shear-induced theory, are explained by Davis [in Ho and Sirkar (eds.), op. cit., pp. 480–505], and Belfort, Davis, and Zydney [*J. Membrane. Sci.,* **96,** 1–58 (1994)]. While not clear-cut, shear-induced diffusion is quite large compared to Brownian diffusion except for those cases with very small particles or very low cross-flow velocity. The enhancement of mass transfer in turbulent-flow microfiltration, a major effect, remains completely empirical.

Fouling Fouling affects MF as it affects all membrane processes. One difference is that the fouling effect caused by deposition of a foulant in the pores or on the surface of the membrane can be confounded by a rearrangement or compression of the solids cake which may form on the membrane surface. Also, the high, open space found in tortuous-pore membranes makes them slower to foul and harder to clean.

Equipment Configuration Since the early days when membrane was available only in flat-sheet form, the variety of offerings of various geometry and fabricated filter component types has grown geometrically. An entire catalog is devoted just to list the devices incorporating membranes whose area ranges from less than 1 cm^2 up to 3 m^2. Microfiltration has grown to maturity selling these relatively small devices. Replacement rather than reuse has long been the custom in MF, and only with later growth of very large applications, such as water, sewage, and corn sweeteners, has long membrane life become an economic necessity on a large scale.

Conventional Designs Designs familiar from other unit operations are also used in microfiltration. Cartridge-filter housings may be fitted with pleated MF membrane making a high-area dead-end membrane filter. Plate-and-frame type devices are furnished with MF sheet stock, and are common in some applications. Capillary bundles with tube-side feed are used for cross-flow applications, and are occasionally used in dead-end flow. A few tubular membranes are used. Spiral-wound modules are becoming increasingly important for process applications where economics are paramount. Belt filters have been made using MF membrane.

Ceramics Ceramic microfilters for commercial applications are almost always employed as tube-side feed multitube monoliths. They are also available as flat sheet, single tubes, discs, and other forms primarily suited to lab use. They are used for a few high-temperature applications, in contact with solvents, and particularly at very high pH.

Cassettes *Cassette* is a term used to describe two different cross-flow membrane devices. The less-common design is a usually large stack of membrane separated by a spacer, with flow moving in parallel across the membrane sheets. This variant is sometimes referred to as a **flat spiral,** since there is some similarity in the way feed and permeate are handled. The more common cassette has long been popular in the pharmaceutical and biotechnical field. It too is a stack of flat-sheet membranes, but the membrane is usually connected so that the feed flows across the membrane elements in series to achieve higher conversion per pass. Their popularity stems from easy direct scale-up from laboratory to plant-scale equipment. Their limitation is that fluid management is inherently very limited and inefficient. Both types of cassette are very compact and capable of automated manufacture.

Representative Process Applications

Pharmaceutical Removal of suspended matter is a frequent application for MF. Processes may be either clarification, in which the main product is a clarified liquid, or solids recovery. Separating cells or their fragments from broth is the most common application. Clarification of the broth in preparation for product recovery is the usual objective, but the primary goal may be recovery of cells. Cross-flow microfiltration competes well with centrifugation, conventional filtration by rotary vacuum filter or filter press and decantation. MF delivers a cleaner permeate, an uncontaminated, concentrated cell product which may be washed in the process, and generally gives high yields. There is no filter-aid disposal problem. Microfiltration has higher capital costs than the other processes, although total cost may be lower. The recovery of penicillin is an example of a process for which cross-flow microfiltration is generally accepted.

Water and Wastewater Microfiltration is beginning to be applied to large-scale potable-water treatment. Its major advantage is positive removal of cryptosporidium and giardia cysts, and its major disadvantage is cost. MF is used in a few large sewage-treatment facilities, where its primary advantage is that it permits a major reduction in the physical size of the facility.

Chemical MF is used in several applications to recover caustic values from cleaning or processing streams. An example is the caustic solution used to clean dairy evaporators, which may be cleaned for reuse by passing it through a microfilter. Significant savings in caustic purchase and disposal costs provide the incentive. Acids are also recovered and reused. Ceramic microfilters are most commonly used in these applications.

Food and Dairy Microfiltration has many applications in the food and dairy industries. An innovative dairy application uses MF membranes to remove bacteria as a nonthermal means of disinfection for milk. A special flow apparatus maintains a carefully controlled transmembrane pressure as the milk flows across the membrane. The concentrate contains the bacteria and spores, as well as any fat. The concentrate may be heat sterilized and recombined with the sterile permeate. In another milk application, some success is reported in separating fat from milk or other dairy streams by cross-flow microfiltration instead of centrifugation. Transmembrane pressure must be kept very low to prevent fat penetration into the membrane. In the food industry, MF membranes are replacing diatomaceous earth filtration in the processing of gelatin. The gelatin is passed with the permeate, but the haze producing components are retained. UF may be used downstream to concentrate the gelatin.

Flow Schemes The outline of batch, semibatch, and stages-in-series is given in the section describing ultrafiltration. Diafiltration is also described there. All these techniques are common in MF, except for stages-in-series, used rarely. MF features uses of special techniques to control transmembrane pressure in some applications. An example is one vendor's device for the microfiltration of milk. In most devices the permeate simply leaves by the nearest exit, but for this application the permeate is pumped through the device in such a way as to duplicate the pressure drop in the concentrate side, thus maintaining a constant transmembrane-pressure driving force. In spite of the low-pressure driving force, the flux is extremely high.

Limitations Some applications which seem ideal for MF, for example the clarification of apple juice, are done with UF instead. The reason is the presence of deformable solids which easily plug and blind an MF membrane. The pores of an ultrafiltration membrane are so small that this plugging does not occur, and high fluxes are maintained. UF can be used because there is no soluble macromolecule in the juice that is desired in the filtrate. There are a few other significant applications where MF seems obvious, but is not used because of deformable particle plugging.

Economics Microfiltration may be the triumph of the Lilliputians; nonetheless, there are a few large-industrial applications. Dextrose plants are very large, and as membrane filtration displaces the precoat filters now standard in the industry, very large membrane microfiltration equipment will be built.

Site Size Most MF processes require a smaller footprint than competing processes. Reduction in total-area requirements are sometimes a decisive economic advantage for MF. It may be apparent that the floor-space costs in a pharmaceutical facility are high, but municipal facilities for water and sewage treatment are often located on expensive real estate, giving MF an opportunity despite its higher costs otherwise.

Large Plants The economics of microfiltration units costing about $\$10^6$ is treated under ultrafiltration. When ceramic membranes are used, the cost optimum may shift energy consumption upward to as much as 10 kWh/m^3.

Disposables For smaller MF applications, short membrane life is a traditional characteristic. In these applications, costs are dominated by the disposables, and an important characteristic of equipment design is the ease, economy, and safety of membrane replacement.

Hygiene and Regulation Almost unique to MF is the influence of regulatory concerns in selection and implementation of a suitable microfilter. Since MF is heavily involved with industries regulated by the Food and Drug Administration, concerns about process stability, consistency of manufacture, virus reduction, pathogen control, and material safety loom far larger than is usually found in other membrane separations.

GAS-SEPARATION MEMBRANES

Process Description Gas-separation membranes separate gases from other gases. Some gas filters, which remove liquids or solids from gases, are microfiltration membranes. Gas membranes generally work because individual gases differ in their solubility and diffusivity through nonporous polymers. A few membranes operate by sieving, Knudsen flow, or chemical complexation.

Selective gas permeation has been known for generations, and the early use of palladium silver-alloy membranes achieved sporadic industrial use. Gas separation on a massive scale was used to separate U^{235} from U^{238} using porous (Knudsen flow) membranes. An upgrade of the membranes at Oak Ridge cost \$1.5 billion. Polymeric membranes became economically viable about 1980, introducing the modern era of gas-separation membranes. H_2 recovery was the first major application, followed quickly by acid gas separation (CO_2/CH_4) and the production of N_2 from air.

The more permeable component is called the *fast gas,* so it is the one enriched in the permeate stream. Permeability through polymers is the product of solubility and diffusivity. The diffusivity of a gas in a membrane is inversely proportional to its kinetic diameter, a value determined from zeolite cage exclusion data (see Table 20-26 after Breck, *Zeolite Molecular Sieves,* Wiley, New York, 1974, p. 636). Tables 20-27, 20-28, and 20-29 provide units conversion factors useful for calculations related to gas-separation membrane systems.

Leading Examples These applications are commercial, some on a very large scale. They illustrate the range of application for gas-separation membranes. Unless otherwise specified, all use polymeric membranes.

Hydrogen Hydrogen recovery was the first large commercial membrane gas separation. Polysulfone fiber membranes became available in 1980 at a time when H_2 needs were rising, and these novel membranes quickly came to dominate the market. Applications include recovery of H_2 from ammonia purge gas, and extraction of H_2 from petroleum cracking streams. Hydrogen once diverted to low-quality fuel use is now recovered to become ammonia, or is used to desulfurize fuel, etc. H_2 is the fast gas.

Carbon Dioxide-Methane Much of the natural gas produced in the world is coproduced with an acid gas, most commonly CO_2 and/or H_2S. While there are many successful processes for separating the gases, membrane separation is a commercially successful competitor, especially for small installations. The economics work best for feeds with very high or very low CH_4 content. Methane is a slow gas; CO_2, H_2S, and H_2O are fast gases.

Oxygen-Nitrogen Because of higher solubility, in many polymers, O_2 is faster than N_2 by a factor of 5. Water is much faster still. Since simple industrial single-stage air compressors provide sufficient pressure to drive an air-separating membrane, moderate purity N_2 (95–99.5%) may be produced in low to moderate quantities quite

TABLE 20-26 Kinetic Diameters for Important Gases

Penetrant	He	H_2	NO	CO_2	Ar	O_2	N_2	CO	CH_4	C_2H_4	Xe	C_3H_8
Kinetic dia, nm	0.26	0.289	0.317	0.33	0.34	0.346	0.364	0.376	0.38	0.39	0.396	0.43

TABLE 20-27 Gas-Permeation Units

Quantity	Engineering units	Literature units	SI units
Permeation rate	Standard cubic feet/minute		kmol/s
Permeation flux	ft^3/ft^2·day	cm/sec (STP)	$kmol/m^2$·s
Permeability	ft^3·ft/ft^2·day·psi	Barrers	kmol/m·s·Pa
Permeance	ft^3/ft^2·day·psi	Barrers/cm	$kmol/m^2$·s·Pa

economically by membrane separation. (Argon is counted as part of the nitrogen.) An O_2-enriched stream is a coproduct, but it is rarely of economic value. The membrane process to produce O_2 as a primary product has a limited market.

Helium Helium is a very fast gas, and may be recovered from natural gas through the use of membranes. More commonly, membranes are used to recover He after it has been used and become diluted.

Gas Dehydration Water is extremely permeable in polymer membranes. Dehydration of air and other gases is a growing membrane application.

Vapor Recovery Organic vapors are recovered from gas streams using highly permeable rubbery polymer membranes which are generally unsuitable for permanent gas separations because of poor selectivity. The high sorption of vapors in these materials makes them ideal for stripping and recovering vapors from gases.

Competing Processes Membranes are not the only way to make these separations, neither are they generally the dominant way. In many applications, membranes compete with cryogenic distillation and with pressure-swing adsorption; in others, physical absorption is the dominant method. The growth rate for membrane capacity is higher than that for any competitor.

Basic Principles of Operation Gas-separation literature often uses nomenclature derived from distillation, a practice that will generally be followed here. L is the molar feed rate, V is the molar permeate rate, R is the molar residue $(L - V)$. Mole fractions of components i, j, in the feed-residue phase will be x_i, x_j . . . and in the permeate phase y_i, y_j. . . . Stage cut, Θ, is permeate volume/feed volume, or V/L.

Basic Equations In "Background and Definitions," the basic equation for gas permeation was derived with the major assumptions noted.

$$J_i \sim N_i = (\rho_i/z)(p_{i,\text{feed}} - p_{i,\text{permeate}}) \tag{20-85}$$

where ρ_i is the permeability of component i through the membrane, J_i is the flux of component i through the membrane for the *partial pressure difference* (Δp) of component i. z is the effective thickness of the membrane. By choosing units appropriately, $J = \rho\Delta p$.

A similar equation may be written for a second component, j, and any additional number of components, employing partial pressures:

$$J_j = (\rho_j/z)(p_{j,\text{feed}} - p_{j,\text{permeate}}) \tag{20-86}$$

The total pressure is the sum of the partial pressures:

$$P_{\text{feed}} = \sum (p_{\text{feed}})_{i,j,...} \tag{20-87}$$

$$P_{\text{permeate}} = \sum (p_{\text{permeate}})_{i,j...} \tag{20-88}$$

For simplicity, consider a two-component system. The volume fraction of a component is

$$x_i = \frac{p_{\text{feed}}}{P_{\text{feed}}} \tag{20-89}$$

$$y_i = \frac{p_{\text{permeate}}}{P_{\text{permeate}}} \tag{20-90}$$

When two species are permeating through a membrane, the ratio of their fluxes can be written following Eqs. (20-85) and (20-86) as:

$$\frac{J_i}{J_j} = \frac{\dfrac{\rho_i}{z}(p_{i,\text{feed}} - p_{i,\text{permeate}})}{\dfrac{\rho_i}{z}(p_{j,\text{feed}} - p_{j,\text{permeate}})} \tag{20-91}$$

Recalling Eq. (20-55), and restating it in the nomenclature for gas membranes:

$$\alpha = \frac{y_i/y_j}{x_i/x_j} \tag{20-92}$$

Defining the pressure ratio $\Phi = P_{\text{feed}}/P_{\text{permeate}}$ and applying Eqs. (20-87) through (20-90) gives:

$$\frac{J_i}{J_j} = \alpha\left[\frac{x_i - (y_i/\Phi)}{x_j - (y_j/\Phi)}\right] \tag{20-93}$$

Combining these equations and rearranging, the permeate composition may be solved explicitly:

$$y_i = \left(\frac{\Phi}{2}\right)\left[x_i + \frac{1}{\Phi} + \frac{1}{\alpha - 1} - \sqrt{\left(x_i + \frac{1}{\Phi} + \frac{1}{\alpha - 1}\right)^2 - \frac{4\alpha x_i}{(\alpha - 1)\Phi}}\right] \tag{20-94}$$

Equation (20-94) gives a permeate concentration as a function of the feed concentration at a stage cut, $\Theta = 0$, To calculate permeate composition as a function of Θ, the equation may be used iteratively if the permeate is unmixed, such as would apply in a test cell. The calculation for real devices must take into account the fact that the driving force is variable due to changes on both sides of the membrane, as partial pressure is a point function, nowhere constant. Using the same caveat, permeation rates may be calculated component by component using Eq. (20-86) and permeance values. For any real device, both concentration and permeation require iterative calculations dependent on module geometry.

Driving Force Gas moves across a membrane in response to a difference in chemical potential. Partial pressure is sufficiently proportional to be used as the variable for design calculations for most gases of interest, but fugacity must be used for CO_2 and usually for H_2 at high pressure. Gas composition changes as a gas passes along a membrane. As the fast gas passes through the membrane, x_i drops. Total pressure on the upstream side of the membrane drops because of frictional losses in the device. Frictional losses on the permeate side will affect the permeate pressure. The partial pressure of a component in the permeate may thus rise rapidly. Permeation rate is a point function, dependent on the difference in partial pressures at a point on the membrane. Many variables affect point partial pressures, among them are membrane structure, module design, and permeate gas-sweep rates. Juxtaposition of feed and permeate is a function of permeator design, and a rapid decline in driving force may result when it is not expected (see "Membrane System Design Features").

TABLE 20-28 Barrer Conversion Factors

Quantity	Multiply	By	To get
Permeability	Barrers	3.348×10^{-19}	kmol/m·s·Pa
Permeability	Barrers	4.810×10^{-8}	ft^3(STP)/ft·psi·day
Permeance	Barrers/cm	3.348×10^{-17}	$kmol/m^2$·s·Pa
Permeance	Barrers/cm	1.466×10^{-6}	ft^3(STP)/ft^2·psi·day

TABLE 20-29 Industry-Specific Gas Measures

Industry-unit	How measured	Cubic feet per pound mole	kmol per mscf
STP, Mscf	1000 ft^3 at 32°F	359.3	1.262
Gas industry, Mscf	1000 ft^3 at 60°F	379.8	1.194
Air industry, Mnsf	1000 ft^3 at 70°F	387.1	1.172

An additional complication may arise in a few cases from the Joule-Thompson effect during expansion of a gas through a membrane changing the temperature. High-pressure CO_2 is an example.

Plasticization Gas solubility in the membrane is one of the factors governing its permeation, but the other factor, diffusivity, is not always independent of solubility. If the solubility of a gas in a polymer is too high, plasticization and swelling result, and the critical structure that controls diffusion selectivity is disrupted. These effects are particularly troublesome with condensable gases, and are most often noticed when the partial pressure of CO_2 or H_2S is high. H_2 and He do not show this effect This problem is well known, but its manifestation is not always immediate.

Limiting Cases Equation (20-94) has two limiting cases for a binary system. First, when $\alpha \gg \phi$. In this case, selectivity is no longer very important.

$$y_i \cong x_i \frac{P_{\text{feed}}}{P_{\text{permeate}}} = x_i \Phi \tag{20-95}$$

Module design is very important for this case, as the high α may result in high permeate partial pressure. An example is the separation of H_2O from air.

Conversely, when $\Phi \gg \alpha$, pressure ratio loses its importance, and permeate composition is:

$$y_i \cong \frac{x_i \alpha}{[1 + x_i(\alpha - 1)]} \tag{20-96}$$

Module design for this case is of lesser importance.

Selectivity and Permeability

State of the Art A desirable gas membrane has high separating power (α) and high permeability to the fast gas, in addition to critical requirements discussed below. The search for an ideal membrane produced copious data on many polymers, neatly summarized by Robeson [*J. Membrane Sci.*, **62,** 165 (1991)]. Plotting log permeability versus log selectivity (α), an "upper bound" is found (see Fig. 20-69) which all the many hundreds of data points fit. The data were taken between 20 and 50°C, generally at 25 or 35°C.

The lower line in Fig. 20-69 shows the upper bound in 1980. Although no breakthrough polymers have been reported in the past few years, it would be surprising if these lines remain the state of the art forever. Performance parameters for polymers used for O_2/N_2 separations are given in Table 20-30.

TABLE 20-30 High-Performance Polymers for O_2/N_2*

Polymer	α (O_2/N_2)	ρ (O_2)—Barrers
Poly (trimethylsilylpropyne)	2.0	4000
Tetrabromo Bis A polycarbonate	7.47	1.36
Poly (tert-butyl acetylene)	3.0	300
Vectra polyester	15.3	0.00046
Poly (triazole)	9.0	1.2
Polypyrrolone	6.5	7.9

*Polymers that are near the upper bound and their characteristics.

The upper-bound line connects discontinuous points, but polymers exist near the bound for separations of interest. Whether these will be available as membranes is a different matter. A useful membrane requires a polymer which can be fabricated into a device having an active layer around 50 nm thick. At this thickness, membrane properties may vary significantly from bulk properties, although not by a factor of 2.

The data reported are *permeabilities,* not *fluxes.* Flux is proportional to permeability/thickness. The separations designer must deal with real membranes, for which thickness is determined by factors outside the designer's control. It is vital that flux data are used in design.

Glassy polymers are significantly overrepresented in the high-selectivity region near the upper bound, and rubbery polymers are overrepresented at the high-permeability end, although the highest-permeability polymer discovered, poly(trimethylsilylpropyne) is glassy. Most of the polymers with interesting properties are noncrystalline. Current membrane-materials research is strongly focused on glassy materials and on attempts to improve diffusivity, as it seems more promising than attempts to increase solubility [Koros, *North Am. Membrane Soc. Sixth Annual Meeting,* Breckenridge, CO (1994)].

Robeson [*J. Membrane Sci.,* **62,** 165 (1991); *Polymer,* **35,** 4970 (1994)] has determined upper-bound lines for many permeant pairs in hundreds of polymers. These lines may be drawn from Eq. (20-97) and the data included in Table 20-31. These values will give ρ_i in

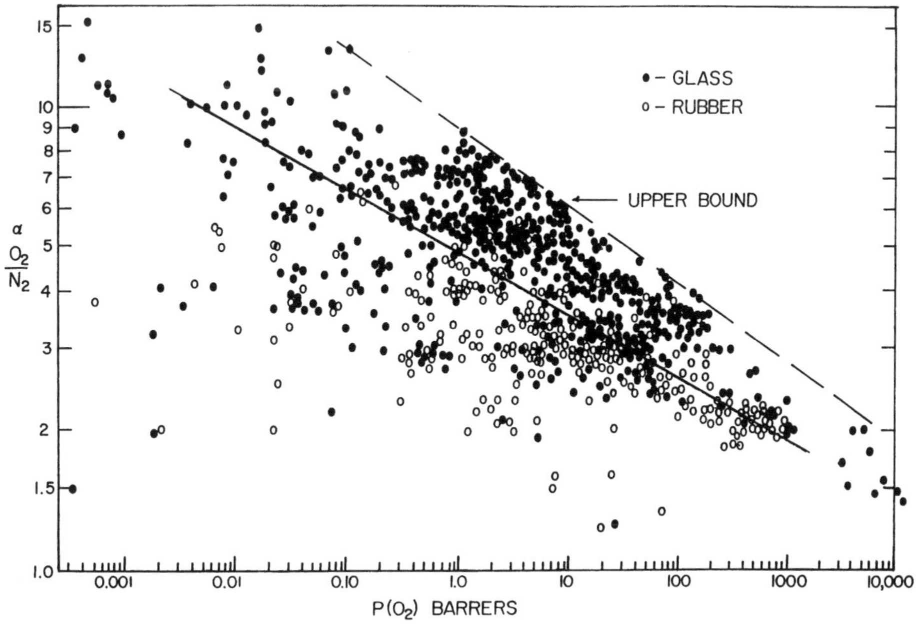

FIG. 20-69 Plot of separation factor versus permeability for many polymers, O_2/N_2. Abscissa—"Fast Gas Permeability, ρ(O_2) Barrers." Ordinate—"Selectivity, α (O_2/N_2)."

TABLE 20-31 Upper-Bound Coordinates for Gas Pairs

Gas pair	log *k*	*m*
He/N_2	4.0969	1.0242
H_2/N_2	4.7236	1.5275
He/CH_4	3.6991	0.7857
H_2/CH_4	4.2672	1.2112
O_2/N_2	5.5902	5.800
He/O_2	3.6628	1.295
H_2/O_2	4.5534	2.277
CO_2/CH_4	6.0309	2.6264
He/H_2	2.9823	4.9535
He/CO_2	2.8482	1.220
H_2/CO_2	3.0792	1.9363

Barrers; α is dimensionless. Robeson [op.cit., (1991); op. cit., (1994)] lists high-performance polymers for most of these gas pairs, like Table 20-31.

$$\log \rho_i = \log k - m \log \alpha_{ij} \qquad (20\text{-}97)$$

Temperature Effects A temperature increase in a polymer membrane permits larger segmental motions in the polymer, producing a dramatic increase in diffusivity. Countering this is a decrease in solubility. It increases the size of the gaps in the polymer matrix, decreasing diffusivity selectivity. The net result is that for a glassy polymer, permeability rises while selectivity declines. For organic permeants in rubbery polymers, this trend is often reversed.

Plasticization and Other Time Effects Most data from the literature, including those presented above are taken from experiments where one gas at a time is tested, with α calculated as a ratio of the two permeabilities. If either gas permeates because of a high-sorption coefficient rather than a high diffusivity, there may be an increase in the permeability of all gases in contact with the membrane. Thus, the α actually found in a real separation may be much lower than that calculated by the simple ratio of permeabilities. The data in the literature do not reliably include the plasticization effect. If present, it results in the sometimes slow relaxation of polymer structure giving a rise in permeability and a dramatic decline in selectivity.

Other Caveats Transport in glassy polymers is different from transport in rubbery ones. In glassy polymers, there are two sites in which sorption may occur, and the literature dealing with dual-mode sorption is voluminous. The simplest case describes behavior when the downstream pressure is zero. It is of great help in understanding the theory but of limited value in practice. There are concerns about permeation of mixtures in glassy polymers with reports of crowding out and competitive sorption. Practical devices are built and operated for many streams, and the complications are often minor. But taking data independently determined for two pure gases and dividing them to obtain α in the absence of other facts is risky.

Gas-Separation Membranes

Organic Organic polymer membranes are the basis for almost all commercial gas-separation activity. Early membranes were cellulose esters and polysulfone. These membranes have a large installed base. New installations are dominated by specialty polymers designed for the purpose, including some polyimides and halogenated polycarbonates. In addition to skinned membranes, composites are made from "designer" polymers, requiring as little as 2 g/1000 m^2. The rapid rise of N_2/O_2 membranes in particular is the result of stunning improvements in product uniformity and quality. A few broken fibers in a 100 m^2 module results in the module's being scrapped.

Caulked Membrane manufacturing defects are unavoidable, and pinholes are particularly deleterious in gas-separation membranes. A very effective remedy is to **caulk** the membrane by applying a highly permeable, very thin topcoat over the finished membrane. While the coating will have poor selectivity, it will plug up the gross leaks while impairing the fast gas permeance only slightly. Unless the as-cast membrane is almost perfect, caulking dramatically improves membrane performance.

Metallic Palladium films pass H_2 readily, especially above 300°C. α for this separation is extremely high, and H_2 produced by purification through certain Pd alloy membranes is uniquely pure. Pd alloys are used to overcome the crystalline instability of pure Pd during heating-cooling cycles. Economics limit this membrane to high-purity applications.

Advanced Materials Experimental membranes have shown remarkable separations between gas pairs such as O_2/N_2 whose kinetic diameters (see Table 20-26) are quite close. Most prominent is the carbon molecular sieve membrane, which operates by ultramicroporous molecular sieving. Preparation of large-scale permeators based on ultramicroporous membranes has proven to be a major challenge.

Catalytic A catalytic-membrane reactor is a combination heterogeneous catalyst and permselective membrane that promotes a reaction, allowing one component to permeate. Many of the reactions studied involve H_2. Membranes are metal (Pd, Ag), nonporous metal oxides, and porous structures of ceramic and glass. Falconer, Noble, and Sperry [in Noble and Stern (eds.), op. cit., pp. 669–709] review status and potential developments.

Membrane System Design Features For the rate process of permeation to occur, there must be a driving force. For gas separations, that force is partial pressure (or fugacity). Since the ratio of the component fluxes determines the separation, the partial pressure of each component at each point is important. There are three ways of driving the process: Either high partial pressure on the feed side (achieved by high total pressure), or low partial pressure on the permeate side, which may be achieved either by vacuum or by introduction of a sweep gas. Both of the permeate options have negative economic implications, and they are less commonly used.

Figure 20-70 shows three of the principal operating modes for gas membranes. A critical issue is the actual partial pressure of permeant at a point on the membrane. Flow arrangements for the permeate are very important in determining the efficiency of the separation, in rough analogy to the importance of arrangements in heat exchangers.

Spiral membranes are the usual way to form flat sheet into modules. They have the characteristic that the feed and the permeate move at

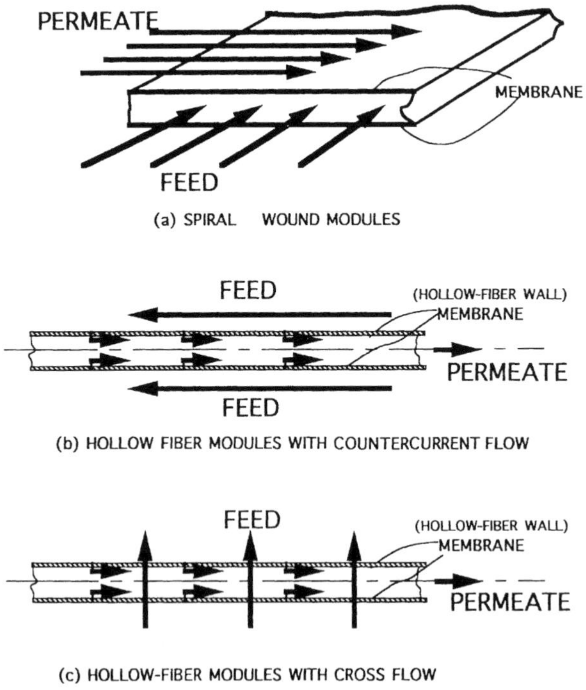

FIG. 20-70 Flow paths in gas permeators. (*Courtesy Elsevier.*)

right angles. Since the membrane is always cast on a porous support, point-permeate values are influenced by the substrate.

Hollow-fiber membranes may be run with shell-side or tube-side feed, cocurrent, countercurrent or in the case of shell-side feed and two end permeate collection, co- and countercurrent. Not shown is the scheme for feed inside the fiber, common practice in lower-pressure separations such as air.

The design of the membrane device will influence whether the membrane is operating near its theoretical limit. Sengupta and Sirkar [in Noble and Stern (eds.), op. cit., pp. 499–552] treat module design thoroughly (including numerical examples for most module configurations) and provide an extensive bibliography.

For a hollow-fiber device running with shell-side feed with the membrane on the outside, Giglia et al. [*Ind. Eng. Chem Research*, **29**, 1239–1248 (1991)] analyzed the effect of membrane-backing porosity on separation efficiency. The application is production of N_2 from air, the desired result being the lowest possible O_2 content in the retentate at a given stage cut. The modules used were operated cocurrent and countercurrent. If the porous-membrane backing prevented the permeate from mixing with the gas adjacent to the membrane, a result approximating cross-flow is expected. For the particular membrane structure used, the experimental result for cocurrent flow was quite close to the calculated cocurrent value, while for countercurrent flow, the experimental data were between the values calculated for the crosscurrent model and the countercurrent model. For a membrane to be commercially useful in this application, the mass transfer on the permeate side must exceed the crosscurrent model.

Air is commonly run with tube-side feed. The permeate is run countercurrent with the separating skin in contact with the permeate. (The feed gas is in contact with the macroporous back side of the membrane.) This configuration has proven to be superior, since the permeate-side mass-transfer problem is reduced to a minimum, and the feed-side mass-transfer problem is not limiting.

Partial Pressure Pinch An example of the limitations of the partial pressure pinch is the dehumidification of air by membrane. While O_2 is the fast gas in air separation, in this application H_2O is faster still. Special dehydration membranes exhibit $\alpha = 20{,}000$. As gas passes down the membrane, the partial pressure of H_2O drops rapidly in the feed. Since the H_2O in the permeate is diluted only by the O_2 and N_2 permeating simultaneously, p_{H_2O} rises rapidly in the permeate. Soon there is no driving force. The commercial solution is to take some of the dry air product and introduce it into the permeate side as a countercurrent sweep gas, to dilute the permeate and lower the H_2O partial pressure. It is in effect the introduction of a leak into the membrane, but it is a controlled leak and it is introduced at the optimum position.

Fouling Industrial streams may contain condensable or reactive components which may coat, solvate, fill the free volume, or react with the membrane. Gases compressed by an oil-lubricated compressor may contain oil, or may be at the water dew point. Materials that will coat or harm the membrane must be removed before the gas is treated. Most membranes require removal of compressor oil. The extremely permeable poly(trimethylsilylpropyne) may not become a practical membrane because it loses its permeability rapidly. Part of the problem is pore collapse, but it seems extremely sensitive to contamination even by diffusion pump oil and gaskets [Robeson, op. cit., (1994)].

A foulinglike problem may occur when condensable vapors are left in the residue. Condensation may result which in the best case results in blinding of the membrane, and in the usual case, destruction of the membrane module. Dew-point considerations must be part of any gas-membrane design exercise.

Modules and Housings Modern gas membranes are packaged either as hollow-fiber bundles or as spiral-wound modules. The former uses extruded hollow fibers. Tube-side feed is preferable, but it is limited to about 1.5 MPa. Higher-pressure applications are usually fed on the shell side. A large industrial permeator contains fibers 400 μm by 200 μm i.d. in a 6-inch shell 10 feet long. Flat-sheet membrane is wound into spirals, with an 8- by 36-inch permeator containing 25 m^2 of membrane. Both types of module are similar to those illustrated in "Background and Definitions." Spiral modules are useful when feed flows are very high and especially in vapor-permeation applications. Otherwise, fiber modules have a large and growing share of the market.

Energy Requirements The thermodynamic minimum energy requirement to separate a metric ton of N_2 from air and compress it to atmospheric pressure at 25°C is independent of separation method, 20.8 MJ or 5.8 kWh. In practice, a cryogenic distillation plant requires twice this energy, and it produces a very pure product as a matter of course. The membrane process requires somewhat more energy than distillation at low purity and much more energy at high purity. Membranes for O_2-reduced air are economical and are a rapidly growing application, but not because of energy efficiency. Figure 20-71 may be used to estimate membrane energy requirements. From the required purity, locate the stage cut on either of the countercurrent flow curves. The compressor work required is calculated using the pressure and the term (product flow rate)/$(1 - \Theta)$. The N_2-rich product is produced at pressure, and the O_2-enriched permeate is vented. For example, if gas is required containing 97 percent inerts, assume the product composition would be 96 percent N_2, 2 percent O_2, 1 percent Ar, and 3 percent O_2, giving a calculated molar mass of 28.2. One tonne would thus contain 35.4 kmol. For 98 percent inerts, Fig. 20-71 shows a stage cut of 67 percent when operating at 690 kPa (abs) and 25°C. Therefore, $35.5/(1 - 0.67) = 108$ kmol of air are required as feed. The adiabatic compression of 108 kmol of air from atmospheric pressure and 300 K (it would subsequently be cooled to $273 + 25 = 298$ K) requires 192 kWh. Assuming 75 percent overall compressor + driver efficiency, 256 kWh are required. For comparison, a very efficient, large N_2 distillation plant would produce 99.99 percent N_2 at 690 kPa for 113 kWh/metric ton. The thermodynamic minimum to separate $(N_2 + Ar)$ and deliver it at the given P and T is 72 kWh.

Economics It is ironic that a great virtue of membranes, their versatility, makes economic optimization of a membrane process very difficult. Designs can be tailored to very specific applications, but each design requires a sophisticated computer program to optimize its costs. Spillman [in Noble and Stern (eds)., op. cit., pp. 589–667] provides an overall review and numerous specific examples including circa 1989 economics.

Rules of Thumb With a few notable exceptions such as H_2 through Pd membranes, membrane separations are not favored when a component is required at high purity. Often, membranes serve these needs by providing a moderate purity product which may be inexpensively upgraded by a subsequent process. Increasing the purity of N_2 by the introduction of H_2 or CH_4 to react with unwanted O_2 is a good example. Unless permeates are recycled, high product purity is accompanied by lower product recovery.

Pressure ratio (Φ) is quite important, but transmembrane ΔP matters as well. Consider the case of a vacuum permeate ($\Phi = \infty$): The membrane area will be an inverse function of P. Φ influences separation and area, transmembrane ΔP influences area.

In a binary separation, the highest purity of integrated permeate occurs at $\Theta = 0$. Purity decreases monotonically until it reaches the feed purity at $\Theta = 1$. In a ternary system, the residue concentration of the gas with the intermediate permeability will reach a maximum at some intermediate stage cut.

Concentration polarization is a significant problem only in vapor separation. There, because the partial pressure of the penetrant is normally low and its solubility in the membrane is high, there can be depletion in the gas phase at the membrane. In other applications it is usually safe to assume bulk gas concentration right up to the membrane.

Another factor to remember is that for $\alpha = 1$, or for $\Phi = 1$, or for $(1/\Theta) = 1$ there is no separation at all. Increasing any of the quantities as defined make for a better separation, but the improvement is diminishing in all cases as the value moves higher. An example of the economic tradeoff between permeability and α is illustrated in Fig. 20-72 where the economics are clearly improved by sacrificing selectivity for flux.

Compression If compression of either feed or permeate is required, it is highly likely that compression capital and operating costs will dominate the economics of the gas-separation process. In some applications, pressure is essentially "free," such as when removing small quantities of CO_2 from natural gas. The gas is often

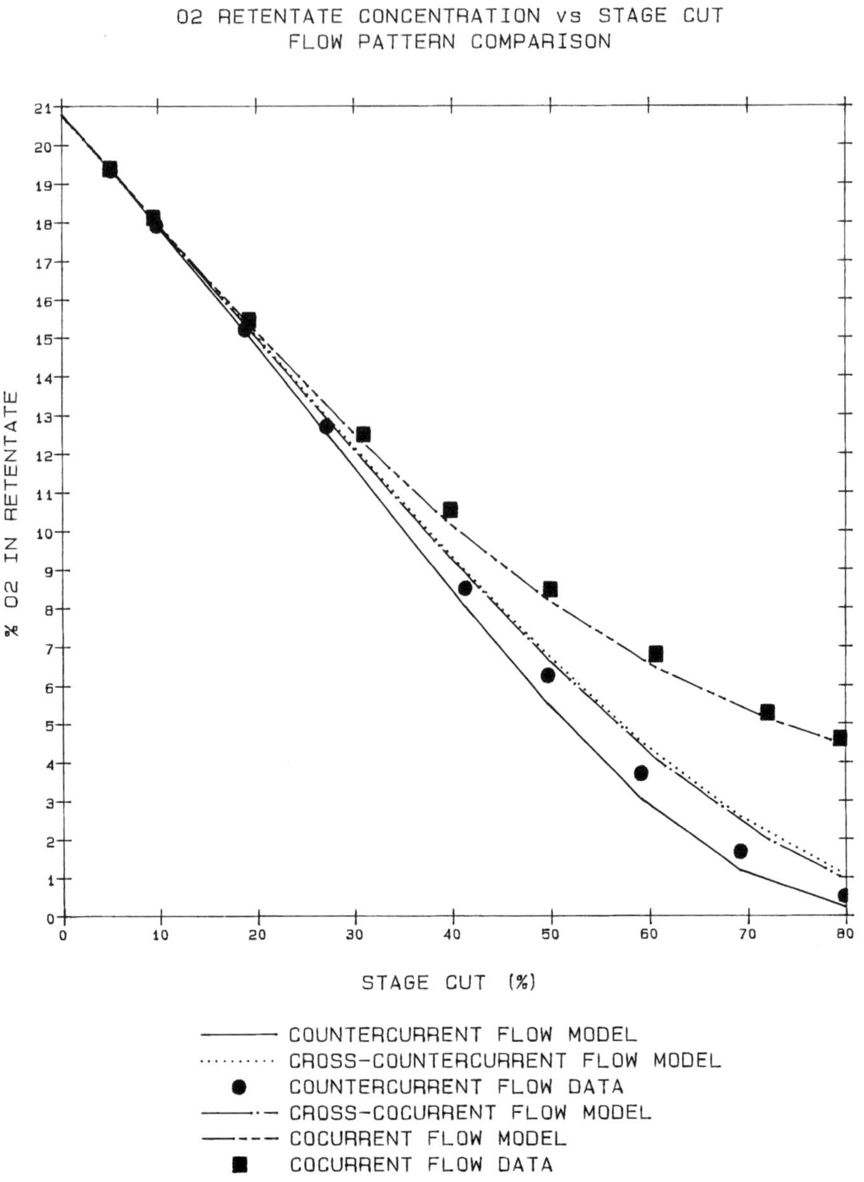

FIG. 20-71 Air fractionation by membrane. O_2 in retentate as a function of feed fraction passed through the membrane (stage cut) showing the different result with changing process paths. Process has shell-side feed at 690 kPa (abs) and 298 K. Module comprised of hollow fibers, diameter 370 μm od × 145 μm id × 1500 mm long. Membrane properties $\alpha = 5.7$ (O_2/N_2), permeance for $O_2 = 3.75 \times 10^{-6}$ Barrer/cm. (*Courtesy Innovative Membrane Systems/ Praxair.*)

produced at pressure, but is compressed for transmission anyway, and since the residue constitutes the product, it continues downstream at pressure. H_2 frequently enters the separation process at pressure, an advantage for membranes. Unlike CH_4, the H_2 is in the permeate, and recompression may be a significant cost. A relative area cure for CO_2/CH_4 is shown in Fig. 20-73. When the permeate is the product (H_2, CO_2) the increasing membrane area shown in Fig. 20-74 is largely the effect of more gas to pass through the membrane, since the curve is based on a constant volume of feed gas, not a constant output of H_2. The facts of life in compressor economics are in painful opposition to the desires of the membrane designer. Pressure ratios higher than six become expensive; vacuum is very expensive, and scale is important. Because of compressor economics, staging membranes with recompression is unusual. Designers can assume that a flow sheet that mixes unlike streams or reduces pressure through a throttling valve will increase cost in most cases.

Product Losses Account must be taken of the product loss, the slow gas in the permeate (such as CH_4), or the fast gas in the residue (such as H_2). Figure 20-75 illustrates the issue for a membrane used to purify natural gas from 93 percent to pipeline quality, 98 percent. In the upper figure, the gas is run through a permeator bank operating as a single stage. For the membrane and module chosen, the permeate contains 63 percent CH_4. By dividing the same membrane area into two stages, two permeates (or more) may be produced, one of which

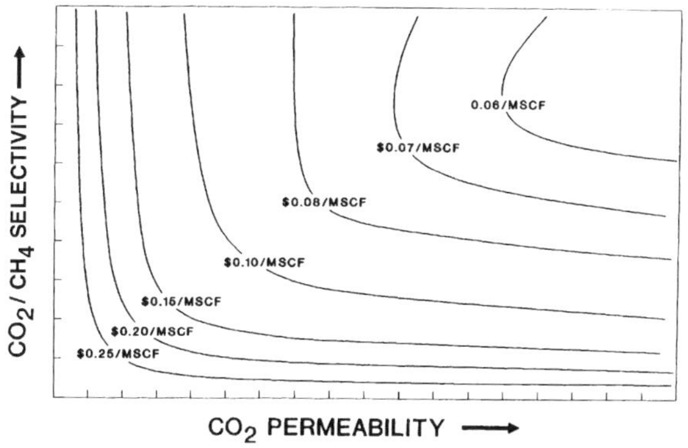

FIG. 20-72 Constant-cost lines as a function of permeability and selectivity for CO_2/CH_4. Cellulose-acetate membrane "mscf" is one thousand standard cubic feet. (*Courtesy W. R. Grace.*)

may have significantly higher economic value than the single mixed permeate. In fact, where CH_4 is involved, another design parameter is the local economic value of various waste streams as fuel.

Membrane Replacement Membrane replacement is a significant cost factor, but membrane life and reliability are now reasonable. Membranes are more susceptible to operating upsets than more traditional equipment, but their field-reliability record in properly engineered, properly maintained installations is good to excellent. In N_2 separations, membrane life is very long.

Competing Technologies The determination of which separation technique is best for a specific application is a dynamic function of advances in membranes and several other technologies. At this writing, very small quantities of pure components are best obtained by purchase of gas in cylinders. For N_2, membranes become competitive at moderate flow rates where purity required is <99.5 percent. At higher flow rates and higher purity, *pressure swing adsorption* (PSA) is better (see Sec. 16: "Adsorption and Ion Exchange"). At still higher volumes, delivered liquid N_2, pipeline N_2, or on-site distillation will be superior. For O_2, membranes have little economic importance. The problem is the economic cost of using vacuum on the permeate side, or the equally unattractive prospect of compressing the feed and operating at low stage cut. For H_2, membranes are dominant when the feed is at high pressure, except for high purity (excepting Pd noted above) and very large volume. Higher purity or lower feed pressure favor PSA. Very high volume favors cryogenic separation. For CH_4, membranes compete at low purity and near pipeline (98 percent) purity, but distillation and absorption are very competitive at large scale and for intermediate CO_2 contamination levels.

Membranes are found as adjuncts to most conventional processes, since their use can improve overall economics in cases where membrane strength coincides with conventional process weakness.

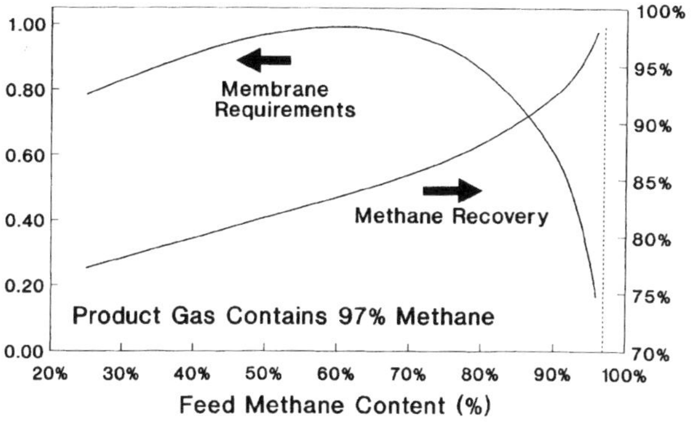

FIG. 20-73 Influence of feed purity on total membrane area when the residue gas at fixed purity is the product. Feed-gas volume is constant. CO_2/CH_4 cellulose-acetate membrane, $\alpha = 21$. (*Courtesy W. R. Grace.*)

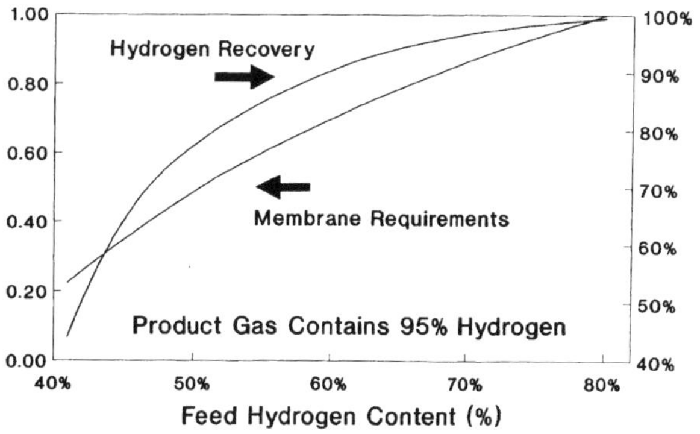

FIG. 20-74 Influence of feed purity on total membrane area when the permeate gas at fixed purity is the product. Feed-gas volume is constant. H_2/CH_4 cellulose-acetate membrane, $\alpha = 45$. (*Courtesy W. R. Grace.*)

PERVAPORATION

Process Description **Pervaporation** is a separation process in which a liquid mixture contacts a nonporous permselective membrane. One component is transported through the membrane preferentially. It evaporates on the downstream side of the membrane leaving as a vapor. The name is a contraction of permeation and evaporation. Permeation is induced by lowering partial pressure of the permeating component, usually by vacuum or occasionally with a sweep gas. The permeate is then condensed or recovered. Thus, three steps are necessary: Sorption of the permeating components into the membrane, diffusive transport across the nonporous membrane, then desorption into the permeate space, with a heat effect. Pervaporation membranes are chosen for high selectivity, and the permeate is often highly purified.

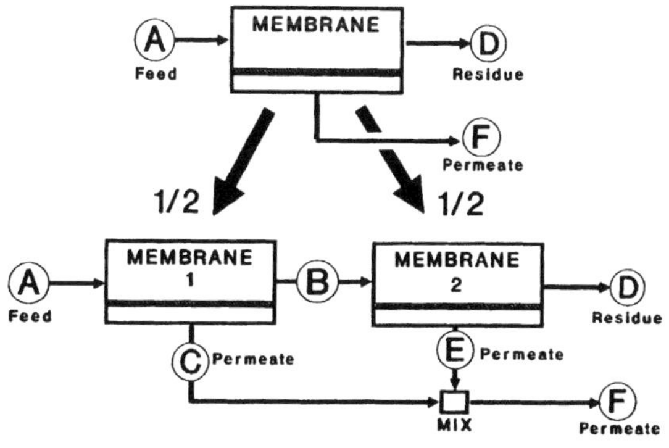

		A	B	C	D	E	F
CO2	%	7.0	3.6	43.9	2.0	27.3	36.6
CH4	%	93.0	96.2	56.1	98.0	72.7	63.4
RATE	Ratio	1.00	0.92	0.08	0.86	0.06	0.14
PRESS	psia	860	860	15	860	15	15

FIG. 20-75 Effect on permeate of dividing a one-stage separation into two equal stages having the same total membrane area. Compositions of A, D, and F are equal for both cases. (*Courtesy W. R. Grace.*)

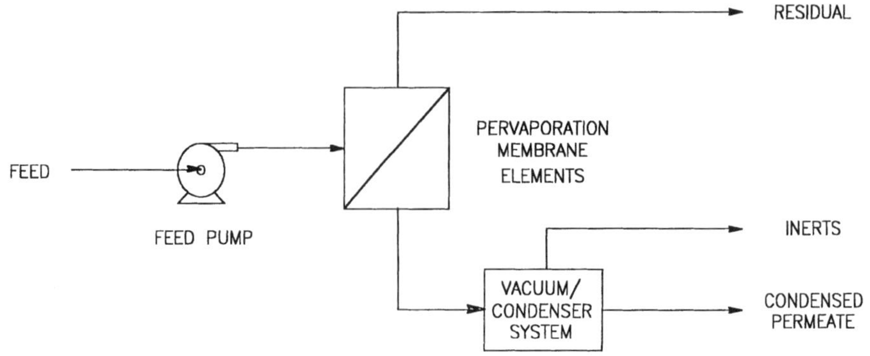

FIG. 20-76 Simplified flow schematic for a pervaporation system. Heated feed enters from left through a feed pump. Heaters in a recirculating feed loop may be required (not shown). Stripped liquid exits at the top of the pervaporation membrane. Vapor exits at the bottom to a condenser. Liquid and noncondensibles are removed under vacuum. (*Courtesy Hoechst Celanese.*)

In the flow schematic (Fig. 20-76), the condenser controls the vapor pressure of the permeating component. The vacuum pump, as shown, pumps both liquid and vapor phases from the condenser. Its major duty is the removal of noncondensibles. Early work in pervaporation focused on organic-organic separations. Many have been demonstrated; few if any have been commercialized. Still, there are prospects for some difficult organic separations.

An important characteristic of pervaporation that distinguishes it from distillation is that it is a rate process, not an equilibrium process. The more permeable component may be the less-volatile component. Pervaporation has its greatest utility in the resolution of azeotropes, as an adjunct to distillation. Selecting a membrane permeable to the minor component is important, since the membrane area required is roughly proportional to the mass of permeate. Thus pervaporation devices for the purification of the ethanol-water azeotrope (95 percent ethanol) are always based on a hydrophilic membrane.

Pervaporation membranes are of two general types. **Hydrophilic membranes** are used to remove water from organic solutions, often from azeotropes. **Hydrophobic membranes** are used to remove organic compounds from water. The important operating characteristics of hydrophobic and hydrophilic membranes differ. Hydrophobic membranes are usually used where the solvent concentration is about 6 percent or less—above this value, other separation methods are usually cheaper unless the flow rate is small. At low-solvent levels the usual membrane (silicone rubber) is not swollen appreciably, and movement of solvent into the membrane makes depletion of solvent in the boundary layer (concentration polarization) an important design problem. Hydrophilic pervaporation membranes operate such that the upstream portion is usually swollen with water, while the downstream face is low in water concentration because it is being depleted by vaporization. Fluxes are low enough (<5 kg/m^2·hr) that boundary layer depletion (liquid side) is not limiting.

The simplifying assumptions that make Fick's law useful for other processes are not valid for pervaporation. The activity gradient across the membrane is far more important than the pressure gradient. Equation (20-98) is generally used to describe the pervaporation process:

$$J_i = -D_i C \frac{d \ln a_i}{dz} \qquad (20\text{-}98)$$

where a_i is the activity coefficient of component i.

This equation is not particularly useful in practice, since it is difficult to quantify the relationship between concentration and activity. The Flory-Huggins theory does not work well with the crosslinked semicrystalline polymers that comprise an important class of pervaporation membranes. Neel (in Noble and Stern, op. cit., pp. 169–176) reviews modifications of the Stefan-Maxwell approach and other equations of state appropriate for the process.

A typical permeant-concentration profile in a pervaporation membrane is shown in Fig. 20-77. The concentration gradient at the permeate (vapor side) of the hydrophilic membrane is usually rate controlling. Therefore the downstream pressure (usually controlled by condenser temperature) is very important for flux and selectivity. Since selectivity is the ratio of fluxes of the components in the feed, as downstream pressure increases, membrane swelling at the permeate interface increases, and the concentration gradient at the permeate interface decreases. The permeate flux drops, and the more swollen membrane is less selective. A rise in permeate pressure may result in a drastic drop in membrane selectivity. This effect is diminished at low water concentrations, where membrane swelling is no longer dominant. In fact, when the water concentration drops far enough, permeate backpressure looses its significance. (See Fig. 20-78.)

For rubbery membranes (hydrophobic), the degree of swelling has less effect on selectivity. Thus the permeate pressure is less critical to the separation, but it is critical to the driving force, thus flux, since the vapor pressure of the organic will be high compared to that of water.

Definitions Following the practice presented under "Gas-Separation Membranes," distillation notation is used. Literature articles often use mass fraction instead of mole fraction, but the substitution of one to the other is easily made.

$$\beta \equiv \frac{y_i}{x_i} \qquad (20\text{-}99)$$

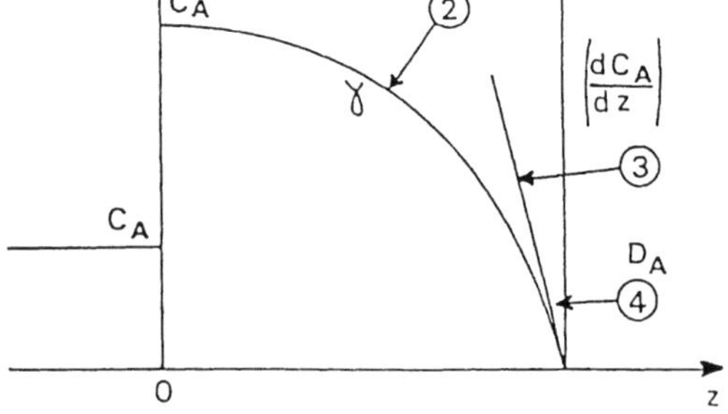

FIG. 20-77 Permeant-concentration profile in a pervaporation membrane. 1—Upstream side (swollen). 2—Convex curvature due to concentration-dependent permeant diffusivity. 3—Downstream concentration gradient. 4—Exit surface of membrane, depleted of permeant, thus unswollen. (*Courtesy Elsevier.*)

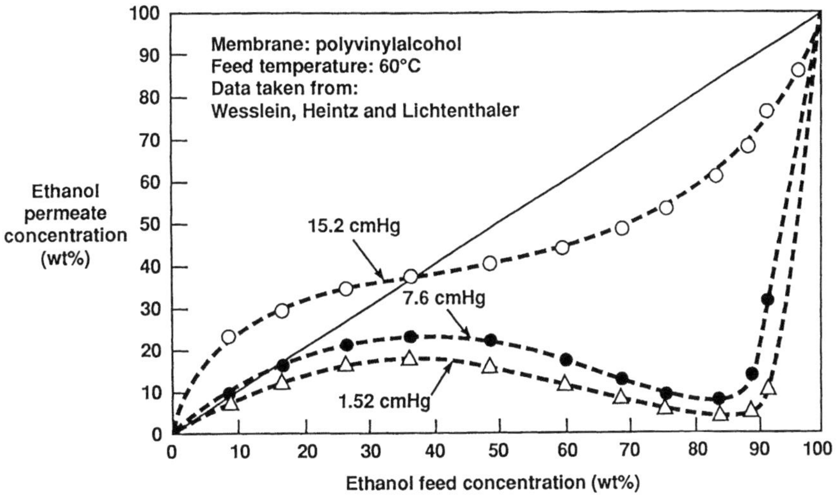

FIG. 20-78 Pervaporation schematic for ethanol-water. The illustration shows the complex behavior for a simple system at three pressures. Only the region above 90 percent is of commercial interest. (*Courtesy Elsevier.*)

where β is the enrichment factor. β is related to α [Eq. (20-92)], the separation factor, by

$$\beta = \frac{\alpha}{1 + (\alpha - 1)x_i} \qquad \alpha = \frac{\beta(1 - x_i)}{1 - x_i\beta} \tag{20-100}$$

α is larger than β, and conveys more meaning when the membrane approaches the ideal. β is preferred in pervaporation because it is easier to use in formulations for cost, yield, and capacity. In fact, note that neither factor is constant, and that both generally change with x_i and temperature.

Operational Factors In industrial use, pervaporation is a continuous-flow single-stage process. Multistage cascade devices are unusual. Pervaporation is usually an adjunct separation, occasionally a principal one. It is used either to break an azeotrope or to concentrate a minor component. Large stand-alone uses may develop in areas where distillation is at a disadvantage, such as with closely boiling components, but these developments, have not yet been made. Notable exceptions occur when a stream is already fairly pure, such as a contaminated water source, or isopropanol from microelectronics fabrication washing containing perhaps 15 percent water.

In continuous flow, the feed will begin with concentration ($x_{i,\text{feed}}$) and be stripped until it reaches some desired ($x_{i,\text{residue}}$). If significant mass is removed from the feed, it will cool, since the general rule for liquids is that the latent heat exceeds the specific heat per °C by two orders of magnitude (more for water). So the membrane has a concentration gradient and a temperature gradient both across it and along it. Pervaporation is a complex multigradient process.

The term β is not constant for some important separations. Even worse, it can exhibit maxima that make analytic treatment difficult. Operating diagrams are often used for preliminary design rather than equations. Because of the very complex behavior in the membrane as concentrations change, all design begins with an experiment. For water removal applications, design equations often mispredict the rate constants as the water content of the feed approaches zero. The estimates tend to be lower than the experimental values, which would lead to overdesign. Therefore it is necessary to obtain experimental data over the entire range of water concentrations encountered in the separation. Once the kinetic data are available, heat transfer and heat capacity are the problem. It is general practice to pilot the separation on a prototype module to measure the changes due to thermal effects. This is particularly true for water as the permeant, given its high latent heat.

For removal of an organic component from water, swelling of the organophillic membrane would result in a higher flux and lower α. At organic levels below about 10 percent, that has not been a major problem. In most applications, boundary-layer mass-transfer limitations become limiting. Pilot data must therefore be taken with hydrodynamic similarity to the module that will be used and the actual organic permeation rate may become limited by the boundary layer. In organic-organic pervaporation, membrane swelling is a major concern.

Vapor Feed A variant on pervaporation is to use vapor, rather than liquid, as a feed. While the resulting process could be classified along with gas-separation membranes, it is customarily regarded as pervaporation.

The residue at the top of a distillation column is a vapor, so there is logic in using it as the feed to a membrane separator. Weighed against the obvious advantage are disadvantages of vapor handling (compressors cost more than pumps) and equipment size to handle the larger vapor volume. When the more volatile component permeates, heat must be added to maintain superheat. The vapor feed technique has been used in a few large installations where the advantages outweigh the disadvantages.

Leading Examples

Dehydration The growing use of isopropanol as a clean-rinse fluid in microelectronics produces significant quantities of an 85–90 percent isopropanol waste. Removing the water and trace contaminants is required before the alcohol can be reused. Pervaporation produces a 99.99 percent alcohol product in one step. It is subsequently polished to remove metals and organics. In Europe, dehydration of ethanol is the largest pervaporation application. For the very large ethanol plants typical of the United States, pervaporation is not competitive with thermally integrated distillation.

Organic from Water An area where pervaporation may become important is in flavors, fragrances, and essential oils. Here, high-value materials with unique properties are recovered from aqueous or alcohol solutions.

Pollution Control Pervaporation is used to reduce the organic loading of a waste stream, thus effecting product recovery and a reduction in waste-treatment costs. An illustration is a waste stream containing 11 percent (wt) *n*-propanol. The residue is stripped to 0.5 percent and 96 percent of the alcohol is recovered in the permeate as a 45 percent solution. This application uses a hydrophobic, rubbery membrane. The residue is sent to a conventional waste-treatment plant.

Pervaporation Membranes Pervaporation has a long history, and many materials have found use in pervaporation experiments. Cellulosic-based materials have given way to polyvinyl alcohol and blends of polyvinyl alcohol and acrylics in commercial water-removing membranes. These membranes are typically solution cast (from water) on ultrafiltration membrane substrates. It is important to have enough crosslinking in the final polymer to avoid dissolution of the membrane in use. A very thin membrane with little penetration into the UF substrate is required. The substrate can be a problem, as it provides significant pressure drop to the vapor passing through it which, as mentioned above, has serious ramifications for flux and separation efficiency. Many new membranes are under development. Ion-exchange membranes, and polymers deposited by or polymerized by plasma, are frequently mentioned in the literature.

Modules Every module design used in other membrane operations has been tried in pervaporation. One unique requirement is for low hydraulic resistance on the permeate side, since permeate pressure is very low (0.1–1 Pa). The rule for near-vacuum operation is the bigger the channel, the better the transport. Another unique need is for heat input. The heat of evaporation comes from the liquid, and intermediate heating is usually necessary. Of course economy is always a factor. Plate-and-frame construction was the first to be used in large installations, and it continues to be quite important. Some smaller plants use spiral-wound modules, and some membranes can be made as capillary bundles. The capillary device with the feed on the inside of the tube has many advantages in principle, such as good vapor-side mass transfer and economical construction, but it is still limited by the availability of membrane in capillary form.

ELECTRODIALYSIS

General References: This section is based on three publications by Heiner Strathmann, to which the interested reader is referred for greater detail [chap. 6, pp. 213–281, in Noble and Stern (eds.), op. cit.; sec. V, pp. 217–262, in Ho and Sirkar, op. cit.; chap. 8, pp. 8-1–8-53, Baker et al., op. cit.].

Process Description *Electrodialysis* (ED) is a membrane separation process in which ionic species are separated from water, macrosolutes, and all uncharged solutes. Ions are induced to move by an electrical potential, and separation is facilitated by ion-exchange membranes. Membranes are highly selective, passing either anions or cations and very little else. The principle of ED is shown in Fig. 20-79.

The feed solution containing ions enters a compartment whose walls are a cation-exchange and an anion-exchange membrane. If the anion-exchange membrane is in the direction of the anode, as shown for the middle feed compartment, anions may pass through that membrane in response to an electrical potential. The cations can likewise move toward the cathode. When the ions arrive in the adjacent compartments, however, their further progress toward the electrodes is prevented by a membrane having the same electrical charge as the ion. The two feed compartments to the left and right of the central compartment are concentrate compartments. Ions entering these two compartments, either in the feed or by passing through a membrane, are retained, either by a same-charged membrane or by the EMF driving the operation. The figure shows two cells (four membranes) between anode and cathode. In an industrial application, a membrane stack can be composed of hundreds of cells, where mobile ions are alternately being depleted and concentrated.

Many related processes use charged membranes and/or EMF. Electrodialytic water dissociation (water splitting), diffusion dialysis, Donnan dialysis, and electrolysis are related processes. **Electrolysis** (chlorine-caustic) is a process of enormous importance much of which is processed through very special membranes.

Leading Examples Electrodialysis has its greatest use in removing salts from brackish water, where feed salinity is around 0.05–0.5 percent. For producing high-purity water, ED can economically reduce solute levels to extremely low levels as a hybrid process in combination with an ion-exchange bed. ED is not economical for the production of potable water from seawater. Paradoxically, it is also used for the concentration of seawater from 3.5 to 20 percent salt. The concentration of monovalent ions and selective removal of divalent ions from seawater uses special membranes. This process is unique to Japan, where by law it is used to produce essentially all of its domestic table salt. ED is very widely used for deashing whey, where the desalted product is a useful food additive, especially for baby food.

Many ED-related processes are practiced on a small scale, or in unique applications. Electrodialysis may be said to do these things well: separate electrolytes from nonelectrolytes and concentrate electrolytes to high levels. It can do this even when the pH is very low. ED does not do well at: removing the last traces of salt (although the hybrid process, electrodeionization, is an exception), running at high pH, tolerating surfactants, or running under conditions where solubility

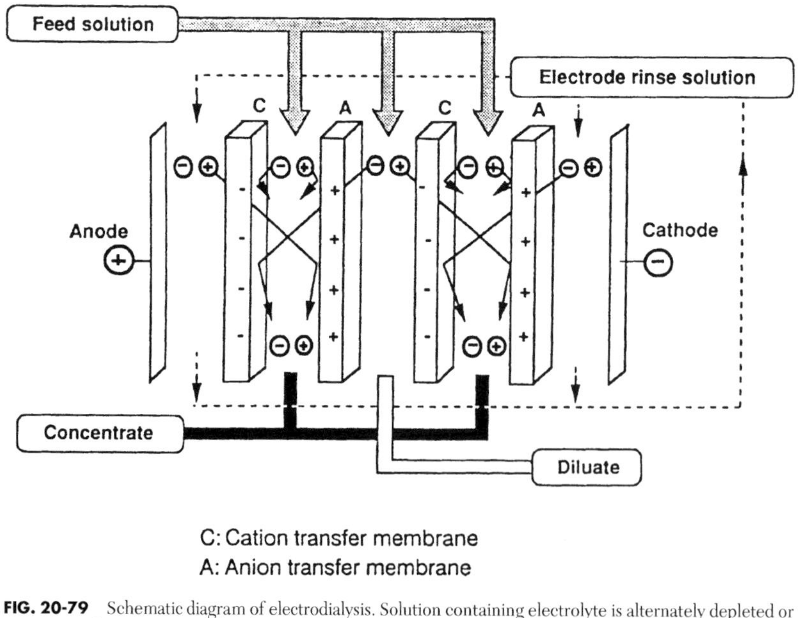

FIG. 20-79 Schematic diagram of electrodialysis. Solution containing electrolyte is alternately depleted or concentrated in response to the electrical field. Feed rates to the concentrate and diluate cells need not be equal. In practice, there would be many cells between electrodes.

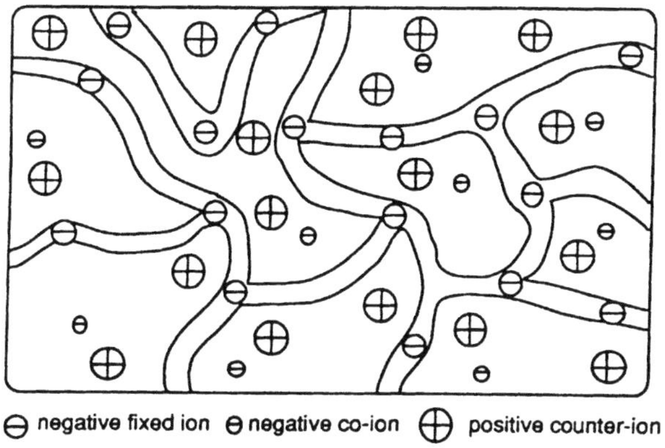

FIG. 20-80 Schematic diagram of a cation-exchange membrane showing the polymer matrix with fixed negative charges and mobile positive counterions. The density of fixed negative charges is sufficient to prevent the passage (exchange) of anions.

limits may be exceeded. Hydroxyl ion and especially hydrogen ion easily permeate both types of ED membrane. Thus, processes that generate a pH gradient across a membrane are limited in their scope.

Water splitting, a closely related process, is useful for reconstituting an acid and a base out of a salt. It is used to reclaim salts produced during neutralization.

Membranes Ion-exchange membranes are highly swollen gels containing polymers with a fixed ionic charge. In the interstices of the polymer are mobile counterions. A schematic diagram of a cation-exchange membrane is depicted in Fig. 20-80.

Figure 20-80 is a schematic diagram of a cation-exchange membrane. The parallel, curved lines represent the polymer matrix composed of an ionic, crosslinked polymer. Shown in the polymer matrix are the fixed negative charges on the polymer, usually from sulfonate groups. The spaces between the polymer matrix are the water-swollen interstices. Positive ions are mobile in this phase, but negative ions are repelled by the negative charge from the fixed charges on the polymer.

In addition to high permselectivity, the membrane must have low-electrical resistance. That means it is conductive to counterions and does not unduly restrict their passage. Physical and chemical stability are also required. Membranes must be mechanically strong and robust, they must not swell or shrink appreciably as ionic strength changes, and they must not wrinkle or deform under thermal stress. In the course of normal use, membranes may be expected to encounter the gamut of pH, so they should be stable from $0 < pH < 14$ and in the presence of oxidants.

Optimization of an ion-exchange membrane involves major trade-offs. Mechanical properties improve with crosslink density, but so does high electrical resistance. High concentration of fixed charges favors low electrical resistance and high selectivity, but it leads to high swelling, thus poor mechanical stability. Membrane developers try to combine stable polymeric backbones with stable ionic functional groups. The polymers are usually hydrophobic and insoluble. Polystyrene is the major polymer used, with polyethylene and polysulfone finding limited application.

Most commercial ion-exchange membranes are homogeneous, produced either by polymerization of functional monomers and cross-linking agents, or by chemical modification of polymers. Many heterogeneous membranes have been prepared both by melting and pressing a mixture of ion-exchange resin and nonfunctional polymer, or by dissolving or dispersing both functional and support resins in a solvent and casting into a membrane. Microheterogeneous membranes have been made by block and graft polymerization.

No membrane and no set of membrane properties has universal applicability. Manufacturers who service multiple applications have a variety of commercial membranes. One firm lists 20 different membranes having a broad spectrum of properties.

Cation-Exchange Membranes Polystyrene copolymerized with divinylbenzene, then sulfonated, is the major building block for cation-exchange membranes. These membranes have reasonable stability and versatility and are highly ionized over most of the pH range. Other chemistries mentioned in the literature include carboxylic acid membranes based on acrylic acid, PO_3^{2-}, HPO_2^-, AsO_3^{2-}, and SeO_3^-. Many specialty membranes have been produced for electrodialysis applications. A notable example is a membrane selective to monovalent cations made by placing a thin coating of positive charge on the cation-exchange membrane. Charge repulsion for polyvalent ions is much higher than that for monovalent ions, but the resistance of the membrane is also higher.

Anion-Exchange Membranes Quaternary amines are the major charged groups in anion-exchange membranes. Polystyrene-divinylbenzene polymers are common carriers for the quaternary amines. The literature mentions other positive groups based on N, P, and S. Anion-exchange membranes are problematic, for the best cations are less robust chemically than their cation exchange counterparts. Since most natural foulants are colloidal polyanions, they adhere preferentially to the anion-exchange membrane, and since the anion-exchange membrane is exposed to higher local pH there is a greater likelihood that precipitates will form there.

Membrane Efficiency The permselectivity of an ion-exchange membrane is the ratio of the transport of electric charge through the membrane by specific ions to the total transport of electrons. Membranes are not strictly semipermeable, for coions are not completely excluded, particularly at higher feed concentrations. For example, the Donnan equilibrium for a univalent salt in dilute solution is:

$$C^M_{Co} = \left(\frac{(C^F_{Co})^2}{C^M_R}\right)\left(\frac{\gamma^F_\pm}{\gamma^M_\pm}\right)^2 \qquad (20\text{-}101)$$

where C^M_{Co} is the concentration of coions (the ions having the same electrical charge as the fixed charges on the membrane); C^F_{Co} is the concentration of such coions in the ambient solution; C^M_R is the concentration of fixed charges in the gel water of the membrane; and $\gamma^F_\pm$ and $\gamma^M_\pm$ are, respectively, the geometric mean of the activity coefficients of the salt ions in the ambient solution and in the membrane. Equation (20-101) is applicable only when $C^F_{Co} << C^M_R$. Since the membrane properties are constant, coion transport rises roughly with the square of concentration.

Process Description Figure 20-81 gives a schematic view of an ED cell pair, showing the salt-concentration profile. In the solution compartment on the left, labeled "Diluate," anions are being attracted to the right by the anode. The high density of fixed cations in membrane "A" is balanced by a high mobile anion concentration within it. Cations move toward the cathode at the left, and there is a similar high mobile cation concentration in the fixed anion membrane "C" separating the depleting compartment from the concentrating compartment further left (not shown). For the anion permeable membrane "A," two boundary layers are shown. They represent depletion in the boundary layer on the left, and an excess in the boundary layer on the right, both due to the concentration polarization effects common to all membrane processes—ions must diffuse through a boundary layer whether they are entering or leaving a membrane. That step must proceed down a concentration gradient.

With every change in ion concentration, there is an electrical effect generated by an electrochemical cell. The anion membrane shown in the middle has three cells associated with it, two caused by the concentration differences in the boundary layers, and one resulting from the concentration difference across the membrane. In addition, there are ohmic resistances for each step, resulting from the E/I resistance through the solution, boundary layers, and the membrane. In solution, current is carried by ions, and their movement produces a friction effect manifested as a resistance. In practical applications, I^2R losses are more important than the power required to move ions to a compartment with a higher concentration.

Transfer of Ions Mass transfer of ions in ED is described by many electrochemical equations. The equations used in practice are empirical. If temperature, the flux of individual components, electroosmotic effects, streaming potential and other indirect effects are

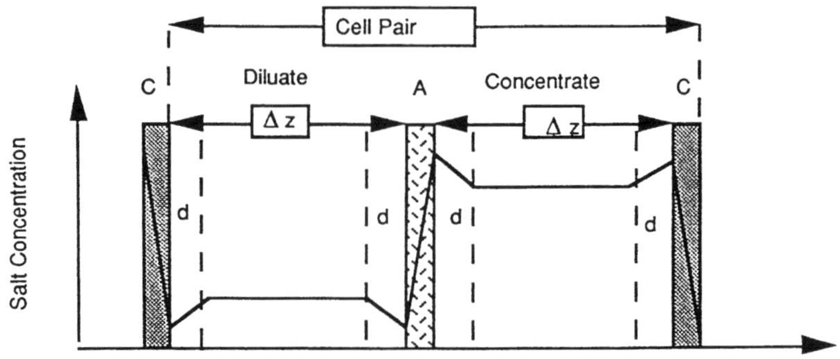

FIG. 20-81 Concentration profile of electrolyte across an operating ED cell. Ion passage through the membrane is much faster than in solution, so ions are enriched or depleted at the cell-solution interface. "d" is the concentration boundary layer. The cell gap Δz should be small. The ion concentration in the membrane proper will be much higher than shown. (*Courtesy Elsevier.*)

minor, an equation good to a reasonable level of approximation is:

$$J_n = C_{mn} U_{mn} \Delta\varphi/\Delta x \tag{20-102}$$

where J is the component flux through the membrane kmol/m^2·s, C_{mn} is the concentration in phase m of component n, kmol/m^3, U_{mn} is the ion mobility of n in m, m^2/v·s, φ is the electrical potential, volts, and x is distance, m.

Equation (20-102) assumes that all mass transport is caused by an electrical potential difference acting only on cations and anions. Assuming the transfer of electrical charges is due to the transfer of ions.

$$i = F \sum_n |z_n J_n| \tag{20-103}$$

where i is the current density, amperes/m^2, F is the Faraday constant and z is the valence. The transport number, T, is the ratio of the current carried by an ion to the current carried by all ions.

$$T_n = \frac{J_n z_n}{\sum_n J_n z_n} \tag{20-104}$$

The transport number is a measure of the permselectivity of a membrane. If, for example, a membrane is devoid of coions, then all current through the membrane is carried by the counterion, and the transport number = 1. The transport numbers for the membrane and the solution are different in practical ED applications.

Concentration Polarization As is shown in the flow schematic, ions are depleted on one side of the membrane and enriched on the other. The ions leaving a membrane diffuse through a boundary layer into the concentrate, so the concentration of ions will be higher at the membrane surface, while the ions entering a membrane diffuse through a boundary layer from the diluate, so the bulk concentration in the diluate must be higher than it is at the membrane. These effects occur because the transport number of counterions in the membrane is always very much higher than their transport number in the solution. Were the transport numbers the same, the boundary layer effects would vanish. This concentration polarization is similar to that experienced in reverse osmosis, except that it has both depletion and enrichment components. The equations governing concentration polarization and depolarization of a membrane are given in the section describing ultrafiltration. The depolarizing strategies used for ED are similar to those employed in other membrane processes, as they involve induced flow past the membrane.

Two basic flow schemes are used: tortuous path flow and sheet flow (Fig. 20-82). Tortuous path spacers are cut to provide a long path between inlet and outlet, providing a relatively long residence time and high velocity past the membrane. The flow channel is open. Sheet flow units have a net spacer separating the membranes. Mass transfer is enhanced either by the spacer or by higher velocity.

Enhanced depolarization requires capital equipment and energy, but it achieves savings in overall capital costs (permits the use of a smaller stack) and energy (permits lower voltage.) The designer's task is to achieve an optimum balance in these requirements. Sheet-flow units have lower capital and operating costs in general, yet both sheet-flow and tortuous-flow units remain competitive. The fact that most ED reversal units (see below) are tortuous flow, and that ED reversal is the dominant technology for water and many waste treatment applications may explain the paradox.

Limiting Current Density As the concentration in the diluate becomes ever smaller, or as the current driving the process is increased, eventually a situation arises in which the concentration of ions at the membranes surrounding the diluate compartment approaches zero. When that occurs, there are insufficient ions to carry additional current, and the cell has reached the limiting current. Forcing the voltage higher results in the dissociation of water at the membrane, giving rise to a dramatic increase in pH due to OH^- ions emerging

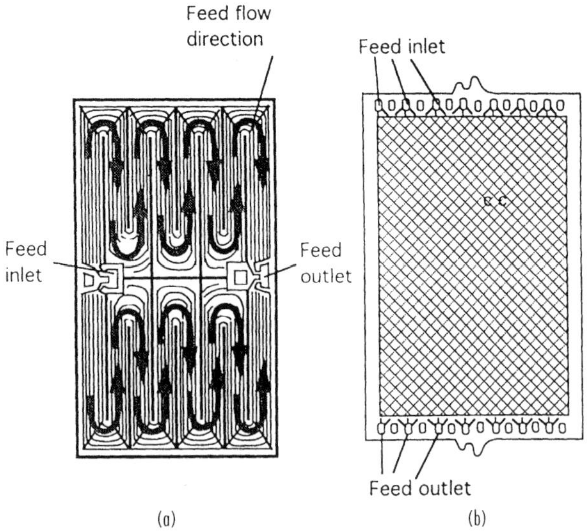

FIG. 20-82 Schematic of two ways to pass solution across an ED membrane. Tortuous flow (left) uses a special spacer to force the solution through a narrow, winding path, raising its velocity, mass transfer, and pressure drop. Sheet feed (right) passes the solution across the plate uniformly, with lower pressure drop and mass transfer. (*Courtesy Elsevier.*)

from the anion-exchange membrane. The high pH promotes precipitation of metal hydroxides and $CaCO_3$ on the membrane surface leading to flow restrictions, poor mass transfer, and subsequent membrane damage. Once a precipitate forms, its presence initiates a vicious cycle fostering the formation of more precipitates. The general expression for the limiting current density is $i_{\text{lim}} \sim (C_{\text{bulk diluate}})$ (diluate velocity)^m, where m is an experimental constant (often 0.6). Thus, the concentration at the membrane is limiting, and at constant current that is proportional to the bulk concentration and the mass-transfer coefficient. Flow in the compartment is laminar but mass transfer is enhanced by the spacer. A normal operating practice is to operate a stack at around 75 percent of the limiting current.

Process Configuration Figure 20-79 shows a basic cell pair. A stack is an assembly of many cell pairs, electrodes, gaskets, and manifolds needed to supply them. An exploded schematic of a portion of a sheet-flow stack is shown in Fig. 20-83.

Gaskets are very important, since they not only keep the streams separated and prevent leaks from the cell, they have the manifolds to conduct feeds, both concentrate and diluate, built into them. No other practical means of feeding the stack is used in the very cramped space required by the need to keep cells thin because the diluate has very low conductivity. The manifolds are formed by aligning holes in membrane and gasket.

The membranes are supported and kept apart by feed spacers. A typical cell gap is 0.5–2 mm. The spacer also helps control solution distribution and enhances mass transfer to the membrane. Given that an industrial stack may have up to 500 cell pairs, assuring uniform flow distribution is a major design requirement.

Since electrodialysis membranes are subject to fouling, it is sometimes necessary to disassemble a stack for cleaning. Ease of reassembly is a feature of ED.

Process Flow The schematic in Fig. 20-79 may imply that the feed rates to the concentrate and diluate compartments are equal. If they are, and the diluate is essentially desalted, the concentrate would leave the process with twice the salt concentration of the feed. A higher ratio is usually desired, so the flow rates of feed for concentrate and feed for diluate can be independently controlled. Since sharply differing flow rates lead to pressure imbalances within the stack, the usual procedure is to recirculate the brine stream using a feed-and-bleed technique. This is usually true for ED reversal plants. Some nonreversal plants use slow flow on the brine side avoiding the recirculating pumps. Diluate production rates are often 10× brine-production rates.

Electrodes No matter how many cells are put in series, there will be electrodes. The more cells, the less the relative importance of the electrodes. The cathode reactions are relatively mild, and depending on the pH:

$$2H^+ + 2e^- \rightarrow H_2$$

$$2H_2O + 2e^- \rightarrow 2OH^- + H_2$$

Anode reactions can be problematic. The anode may dissolve or be oxidized. Or, depending on the pH and the chloride concentration:

$$2H_2O \rightarrow O_2 + 4H^+ + 4e^-$$

$$4OH^- \rightarrow O_2 + 2H_2O + 4e^-$$

$$2Cl^- \rightarrow Cl_2 + 2e^-$$

Dissolution of metal is avoided by selecting a resistant material such as Pt, Pt coated on Ti, or Pt on Nb. Base metals are sometimes used, as are graphite electrodes.

Electrode isolation is practiced to minimize chlorine production and to reduce fouling. A flush solution free of chlorides or with reduced pH is used to bathe the electrodes in some plants. Further information on electrodes may be found in a work by David ["Electrodialysis," pp. 496–499, in Porter (ed.), op. cit.].

Peripheral Components In addition to the stack, a power supply, pumps for diluate and concentrate, instrumentation, tanks for cleaning, and other peripherals are required. Safety devices are mandatory given the dangers posed by electricity, hydrogen, and chlorine.

Pretreatment Feed water is pretreated to remove gross objects that could plug the stack. Additives that inhibit the formation of scale, frequently acid, may be introduced into the feed.

Electrodialysis Reversal Two basic operating modes for ED are used in large-scale installations. **Unidirectional operation** is the mode described above in the general explanation of the process. The electrodes maintain their polarity and the ions always move in a constant direction. **ED reversal** is an intermittent process in which the polarity in the stack is reversed periodically. The interval may be from several minutes to several hours. When the polarity is reversed, the identity of compartments is also reversed, and diluate compartments become concentrate compartments and vice versa. The scheme requires instruments and valves to redirect flows appropriately after a reversal. The advantages that often justify the cost are a major reduction in membrane scaling and fouling, a reduction in feed additives required to prevent scaling, and less frequent stack-cleaning requirements.

Water Splitting A modified electrodialysis arrangement is used as a means of regenerating an acid and a base from a corresponding salt. For instance, NaCl may be used to produce NaOH and HCl. Water splitting is a viable alternative to disposal where a salt is produced by neutralization of an acid or base. Other potential applications include the recovery of organic acids from their salts and the treating of effluents from stack gas scrubbers. The new component required is a bipolar membrane, a membrane that splits water into H^+ and OH^-. At its simplest, a bipolar membrane may be prepared by laminating a cation and an anion membrane. In the absence of mobile ions, water sorbed in the membrane splits into its components when a sufficient electrical gradient is applied. The intimate contact of the two membranes minimizes the problem of the low ionic conductivity of ion-depleted water. As the water is split, replacement water readily

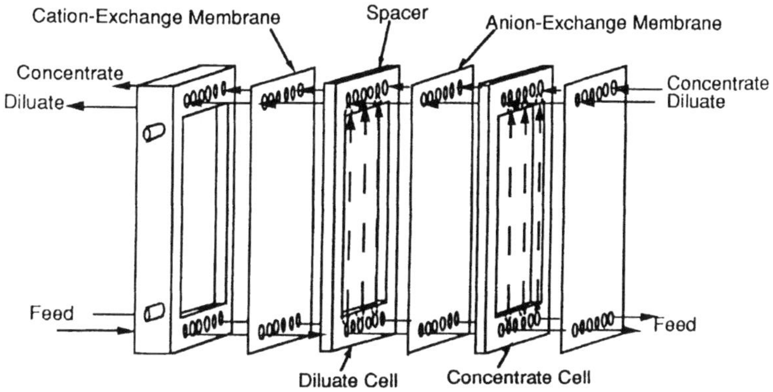

FIG. 20-83 Exploded view of a sheet-feed ED stack. Manifolds are built into the membranes and spacers as the practical way to maintain a narrow cell gap. (*Courtesy Elsevier.*)

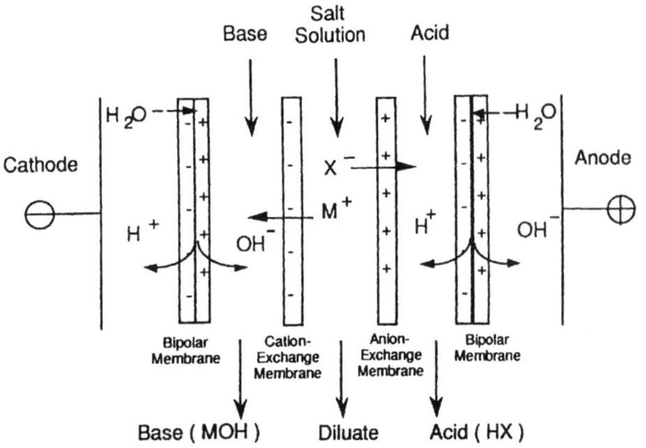

FIG. 20-84 Electrodialysis water dissociation (water splitting) membrane inserted into an ED stack. Starting with a salt, the device generates the corresponding acid and base by supplying H^+ and OH^- from the dissociation of water in a bipolar membrane. (*Courtesy Elsevier.*)

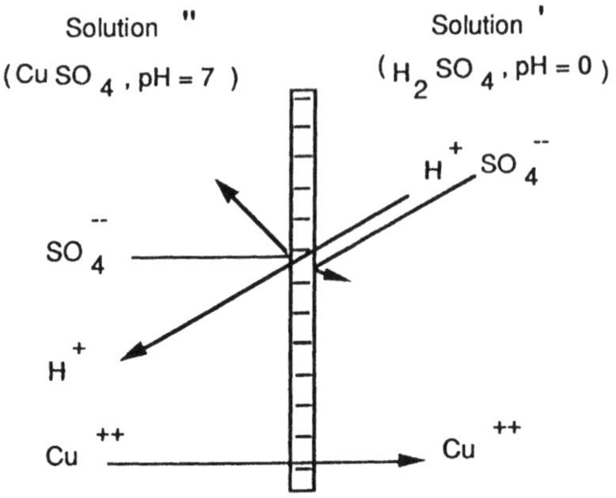

FIG. 20-85 Schematic of Donnan dialysis using a cation-exchange membrane. Cu^{2+} is "pumped" from the lower concentration on the left to a higher concentration on the right maintaining electrical neutrality accompanying the diffusion of H^+ from a low pH on the right to a higher pH on the left. The membrane's fixed negative charges prevent mobile anions from participating in the process. (*Courtesy Elsevier.*)

diffuses from the surrounding solution. Properly configured, the process is energy efficient.

A schematic of the production of acid and base by electrodialytic water dissociation is shown in Fig. 20-84. The bipolar membrane is inserted in the ED stack as shown. Salt is fed into the center compartment, and base and acid are produced in the adjacent compartments. The bipolar membrane is placed so that the cations are paired with OH^- ions and the anions are paired with H^+. Neither salt ion penetrates the bipolar membrane. As is true with conventional electrodialysis, many cells may be stacked between the anode and the cathode.

If recovery of both acid and base is unnecessary, one membrane is left out. For example, in the recovery of a weak acid from its salt, the anion-exchange membrane may be omitted. The process limitations relate to the efficiency of the membranes, and to the propensity for H^+ and OH^- to migrate through membranes of like fixed charge, limiting the attainable concentrations of acid and base to 3–5 N. The problem is at its worst for HCl and least troublesome for organic acids. Ion leakage limits the quality of the products, and the regenerated acids and bases are not of high enough quality to use in regenerating a mixed-bed ion-exchange resin.

Diffusion Dialysis The propensity of H^+ and OH^- to penetrate membranes is useful in diffusion dialysis. An anion-exchange membrane will block the passage of metal cations while passing hydrogen ions. This process uses special ion-exchange membranes, but does not employ an applied electric current.

As an example, in the aircraft industry heavy-aluminum sections are shaped as airfoils, then masked. The areas where the metal is not required to be strong are then unmasked and exposed to NaOH to etch away unneeded metal for weight reduction. Sodium aluminate is generated, a potential waste problem. Cation-exchange membranes leak OH^- by a poorly understood mechanism that is not simply the transport of OH^- with its waters of hydration. The aluminate anion is retained in the feed stream while the caustic values pass through. NaOH recovery is high, because all the Na^+ participates in the driving force. There is considerable passage of water due to the osmotic pressure difference as well. This scheme operates efficiently only because aluminum hydroxide forms highly supersaturated solutions. Hydroxide precipitation within the apparatus is reported to be a minor problem. $Al(OH)_3$ is precipitated in a downstream crystallizer, and is reported to be of high quality.

Donnan Dialysis Another nonelectrical process using ED membranes is used to exchange ions between two solutions. The common application is to use H^+ to drive a cation from a dilute compartment to a concentrated one. A schematic is shown in Fig. 20-85. In the right compartment, the pH is 0, thus the H^+ concentration is 10^7 higher than in the pH 7 compartment on the left. H^+ diffuses leftward, creating an electrical imbalance that can only be satisfied by a cation diffusing rightward through the cation-selective membrane. By this scheme, Cu^{++} can be "pumped" from left to right against a significant concentration difference.

Electrodialysis-Moderated Ion Exchange The production of ultrapure water is facilitated by incorporating a mixed-bed ion-exchange resin between the membranes of an ion-exchange stack. Already pure water is passed through the bed, while an electric current is passed through the stack. Provided the ion-exchange beads are in contact with each other and with the membranes, an electrical current can pass through the bed even though the conductivity of the very pure water is quite low. In passing, the current conducts any ions present into adjacent compartments, simultaneously and continuously regenerating the resin in situ.

Energy Requirements The thermodynamic limit on energy is the ideal energy needed to move water from a saline solution to a pure phase. The theoretical minimum energy is given by:

$$\Delta G = RT \ln (a/a_s) \tag{20-105}$$

where ΔG is the Gibbs free energy required to move one mole of water from a solution, a is the activity of pure water ($\equiv 1$), and a_s is the activity of water in the salt solution. In a solution, the activity of water is approximately equal to the molar fraction of water in the solution. So that approximate activity is

$$a_s = \frac{n_s}{n_s + \nu \cdot s_s} \qquad \frac{1}{a_s} = 1 + \frac{\nu \cdot s_s}{n_s} \tag{20-106}$$

where n_s is the number of moles of water in the salt solution, ν is the number of atoms in the salt molecule (2 for NaCl, 3 for $CaCl_2$) and s_s is the number of moles of salt in the salt solution. The ratio of moles of salt in the salt solution to the number of moles of water in the salt solution is a very small number for a dilute solution. This permits using the approximation $\ln (1 + x) = x$, when x is of the magnitude 0.01, making this an applicable approximation for saline water. That permits rewriting Eq. (20-105) as

$$\Delta G = \nu RT(s_s/n_s) \tag{20-107}$$

where ΔG is still the free energy required to move one mole of water from the saline solution to the pure water compartment.

The conditions utilized in the above development of minimum energy are not sufficient to describe electrodialysis. In addition to the desalination of water, salt is moved from a saline feed to a more concentrated compartment. That free-energy change must be added to the free energy given in Eq. (20-107), which describes the movement of water from salt solution, the reverse of the actions in the diluate compartment (but having equal free energy). *Schaffer & Mintz* develop that change, and after solving the appropriate material balances, they arrive at a practical simplified equation for a monovalent ion salt, where activities may be approximated by concentrations:

$$\Delta G = RT(C_f - C_d)\left(\frac{\ln C_{fc}}{C_{fc}-1} - \frac{\ln C_{fd}}{C_{fd}-1}\right);\ C_{fc} = \frac{C_f}{C_c};\ C_{fd} = \frac{C_f}{C_d} \qquad (20\text{-}108)$$

where C_f is the concentration of ions in the feed, C_d is the concentration in the diluate, and C_c is the concentration in the concentrate, all in kmol/m^3. When $C_{fc} \rightarrow 1$ and $C_{fd} \rightarrow$ infinity, the operation is one approximating the movement of salt from an initial concentration into an unlimited reservoir of concentrate, while the diluate becomes pure. This implies that the concentrate remains at a constant salt concentration. In that case, Eq. (20-108) reduces to $RT(C_f - C_d)$. As a numerical example of Eq. (20-108) consider the desalting of a feed with initial concentration 0.05 M to 0.005 M, roughly approximating the production of drinking water from a saline feed. If 10 ℓ of product are produced for every 3 ℓ of concentrate, the concentrate leaves the process at 0.2 M. The energy calculated from Eq. (20-108) is 0.067 kWh/m^3 at 25°C. If the concentrate flow is infinite, $C_c = 0.05$ M, and the energy decreases to 0.031 kWh/m^3.

This minimum energy is that required to move ions only, and that energy will be proportional to the ionic concentration in the feed. It assumes that all resistances are zero, and that there is no polarization. In a real stack, there are several other important energy dissipaters. One is overcoming the electrical resistances in the many components. Another is the energy needed to pump solution through the stack to reduce polarization and to remove products. Either pumping or desalting energy may be dominant in a working stack.

Energy Not Transporting Ions Not all current flowing in an electrodialysis stack is the result of the transport of the intended ions. Current paths that may be insignificant, minor, or significant include electrical leakage through the brine manifolds and gaskets, and transport of co-ions through a membrane. A related indirect loss of current is water transport through a membrane either by osmosis or with solvated ions, representing a loss of product, thus requiring increased current.

Pump Energy Requirements If there is no forced convection within the cells, the polarization limits the current density to a very uneconomic level. Conversely, if the circulation rate is too high, the energy inputs to the pumps will dominate the energy consumption of the process. Furthermore, supplying mechanical energy to the cells raises the pressure in the cells, and raises the pressure imbalance between portions of the stack, thus the requirements of the confining gear and the gaskets. Also, cell plumbing is a design problem made more difficult by high circulation rates.

A rule of thumb for a modern ED stack is that the pumping energy is roughly 0.5 kWh/m^3, about the same as is required to remove 1700 mg/ℓ dissolved salts.

Equipment and Economics A very large electrodialysis plant would produce 500 ℓ/s of desalted water. A rather typical plant was built in 1993 to process 4700 m^3/day (54.4 ℓ/s). Capital costs for this plant, running on low-salinity brackish feed were $1,210,000 for all the process equipment, including pumps, membranes, instrumentation, and so on. Building and site preparation cost an additional $600,000. The building footprint is 300 m^2. For plants above a threshold level of about 40 m^3/day, process-equipment costs usually scale at around the 0.7 power, not too different from other process equipment. On this basis, process equipment (excluding the building) for a 2000 m^3/day plant would have a 1993 predicted cost of $665,000.

The greatest operating-cost component, and the most highly variable, is the charge to amortize the capital. Many industrial firms use capital charges in excess of 30 percent. Some municipalities assign long amortization periods and low-interest rates, reflecting their cost of capital. Including buildings and site preparation, the range of capital charges assignable to 1000 m^3 of product is $90 to $350.

On the basis of 1000 m^3 of product water, the operating cost elements (as shown in Table 20-32) are anticipated to be:

TABLE 20-32 Electrodialysis Operating Costs

$ 66	Membrane-replacement cost (assuming seven-year life)
32	Plant power
16	Filters and pretreatment chemicals
11	Labor
8	Maintenance
$133	Total

These items are highly site specific. Power cost is low because the salinity removed by the selected plant is low. The quality of the feed water, its salinity, turbidity, and concentration of problematic ionic and fouling solutes, is a major variable in pretreatment and in conversion.

SELECTION OF BIOCHEMICAL SEPARATION PROCESSES

GENERAL REFERENCES: Ahuja (ed.), *Handbook of Bioseparations*, Academic Press, London, 2000. Albertsson, *Partition of Cell Particles and Macromolecules*, 3d ed., Wiley, New York, 1986. Belter, Cussler, and Hu, *Bioseparations*, Wiley Interscience, New York, 1988. Cooney and Humphrey (eds.), *Comprehensive Biotechnology*, vol. 2, Pergamon, Oxford, 1985. Flickinger and Drew (eds.), *The Encyclopedia of Bioprocess Technology: Fermentation, Biocatalysis, and Bioseparations*, Wiley, New York, 1999. Harrison, Todd, Scott, and Petrides, *Bioseparations Science and Engineering*, Oxford University Press, Oxford, 2003. Hatti-Kaul and Mattiasson (eds.), *Isolation and Purification of Proteins*, Marcel Dekker, New York, 2003. Janson and Ryden (eds.), *Protein Purification*, VCH, New York, 1989. Ladisch, *Bioseparations Engineering: Principles, Practice, and Economics*, Wiley, New York, 2001. Scopes, *Protein Purification*, 2d ed., Springer-Verlag, New York, 1987. Stephanopoulos (ed.), *Biotechnology*, 2d ed., vol. 3, VCH, Weinheim, 1993. Subramanian (ed.), *Bioseparation and Bioprocesss—A Handbook*, Wiley, New York, 1998. Zaslavsky, *Aqueous Two-Phase Partitioning—Physical Chemistry and Bioanalytical Applications*, Marcel Dekker, New York, 1995.

GENERAL BACKGROUND

The biochemical industry derives its products from two primary sources. Natural products are yielded by plants, animal tissue, and fluids, and they are obtained via fermentation from bacteria, molds, and fungi and from mammalian cells. Products can also be obtained by recombinant methods through the insertion of foreign DNA directly into the hosting microorganism to allow overproduction of the product in this unnatural environment. The range of bioproducts is enormous, and the media in which they are produced are generally complex and ill-defined, containing many unwanted materials in addition to the desired product. The product is almost invariably at low concentration to start with. The goals of downstream processing operations include removal of these unwanted impurities, bulk-volume reduction with concomitant concentration of the desired product, and, for protein products, transfer of the protein to an environment where it will be stable and active, ready for its intended application. This always requires a multistage process consisting of multiple-unit operations. A general strategy for downstream processing of biological materials and the types of operations that may be used in the different steps is shown in Fig. 20-86 [see also Ho, in M. R. Ladisch et al. (eds.), *Protein Purification from Molecular Mechanisms to Large-Scale Processes, ACS Symp. Ser.*, **427,** ACS, Washington, D.C. (1990), pp. 14–34]. Low-molecular-weight products, generally secondary metabolites such as alcohols, carboxylic and amino acids, antibiotics, and vitamins, can be recovered by using many of the standard operations such as liquid-liquid partitioning, adsorption, and ion exchange, described elsewhere in this handbook. Biofuel molecules also belong

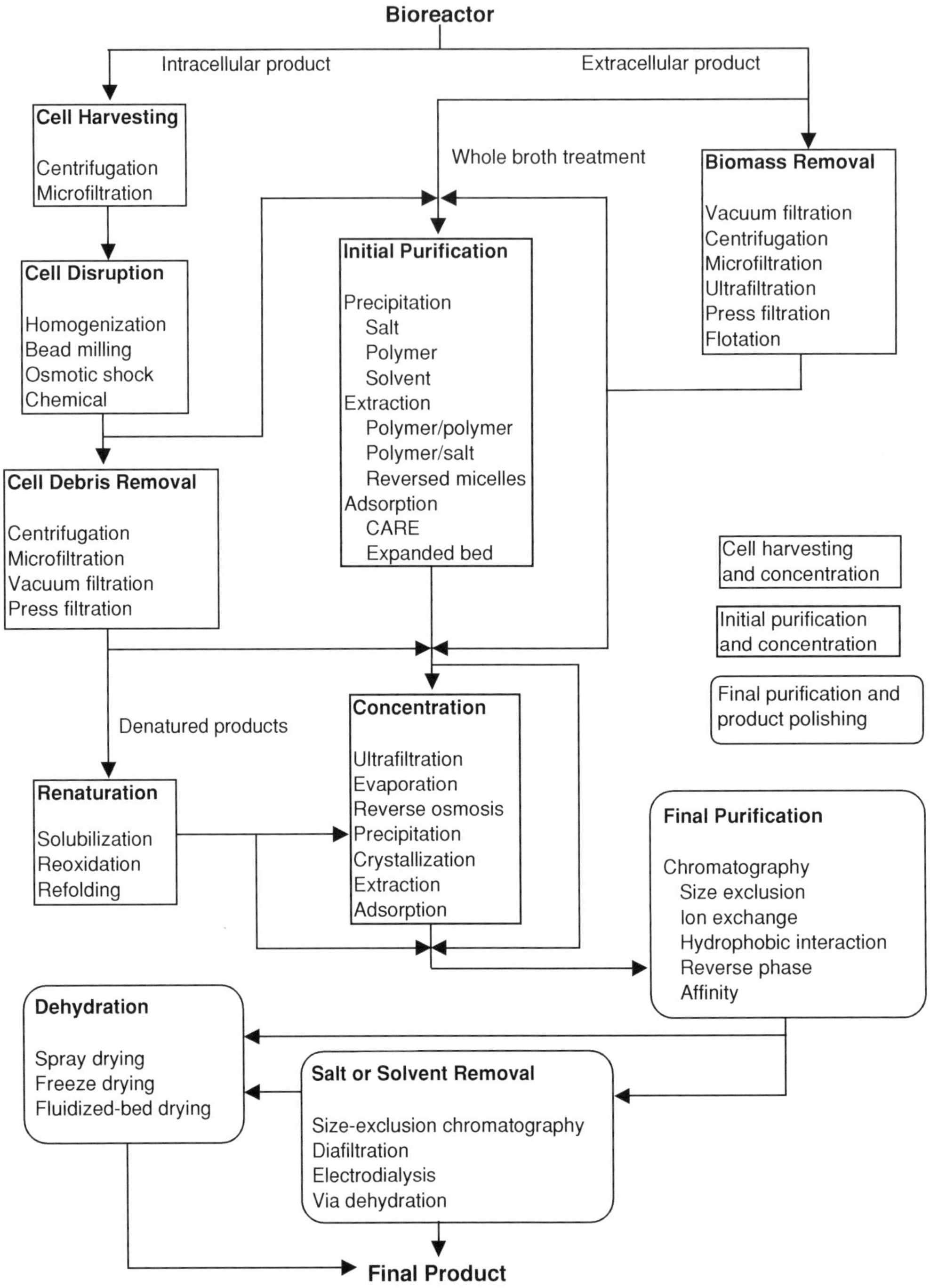

FIG. 20-86 General stages in downstream processing for protein production indicating representative types of unit operations used at each stage.

to this category. They are becoming more and more important. Less expensive and high-throughput unit operations are needed to make a biorefinery process economically feasible. Proteins require special attention, however, as they are sufficiently more complex, their function depending on the integrity of a delicate three-dimensional tertiary structure that can be disrupted if the protein is not handled correctly. For this reason, this section focuses primarily on protein separations.

Techniques used in bioseparations depend on the nature of the product (i.e., the unique properties and characteristics which provide a "handle" for the separation) and on its state (i.e., whether soluble or insoluble, intra- or extracellular, etc.). All early isolation and recovery steps remove whole cells, cellular debris, suspended solids, and colloidal particles; concentrate the product; and in many cases, achieve some degree of purification, all the while maintaining high yield. For intracellular compounds, the initial harvesting of the cells is important for their concentration prior to release of the product. Following this phase, a range of purification steps are employed to remove the remaining impurities and enhance the product purity; this purification phase, in turn, is followed by polishing steps to remove the last traces of contaminating components and process-related additions (e.g., buffer salts, detergents) and to prepare the product for storage and/or distribution. The prevention and/or avoidance of contamination is another important goal of downstream processing. Therapeutic proteins require very high purities that cannot be measured by weight percentages alone. To avoid potential harmful effects in humans, the levels of pyrogens and microbial and viral contaminates in final products must meet stringent safety and regulatory requirements.

Even for good yields of 80 to 95 percent per step, the overall yield can be poor for any process that requires a large number of steps. Thus, careful consideration must be given to optimization of the process in terms of both the unit operations themselves and their sequencing. A low-yield step should be replaced if improvement is unattainable to eliminate its impact on the overall yield. It is usually desirable to reduce the process volume early in the downstream processing, and to remove any components that can be removed fairly easily (particulates, small solutes, large aggregates, nucleic acids, etc.), so as not to overly burden the more refined separation processes downstream. Possible shear and temperature damage, and deactivation by endogenous proteases, must be considered in the selection of separation processes. Protein stability in downstream processing was discussed in depth by Hejnaes et al. [in Subramanian (ed.), vol. 2, op. cit., pp. 31–66].

INITIAL PRODUCT HARVEST AND CONCENTRATION

The initial processing steps are determined to a large extent by the location of the product species, and they generally consist of cell/broth separation and/or cell debris removal. For products retained within the biomass during production, it is first necessary to concentrate the cell suspension before homogenization or chemical treatment to release the product. Clarification to remove the suspended solids is the process goal at this stage.

Regardless of the location of the protein and its state, cell separation needs to be inexpensive, simple, and reliable, as large amounts of fermentation-broth dilute in the desired product may be handled. The objectives are to obtain a well-clarified supernatant and solids of maximum dryness, avoiding contamination by using a contained operation. Centrifugation or crossflow filtration is typically used for cell separation, and both unit operations can be run in a continuous-flow mode [Datar and Rosen, in Stephanopoulos (ed.), op. cit., pp. 369–503]. In recent years, expanded-bed adsorption has become an alternative. It combines broth clarification and adsorption separation in a single step.

Intracellular products can be present either as folded, soluble proteins or as dense masses of unfolded protein (inclusion bodies). For these products, it is first necessary to concentrate the cell suspension before effecting release of the product. Filtration can result in a suspension of cells that can be of any desired concentration up to 15 to 17 percent and that can be diafiltered into the desired buffer system. In contrast, the cell slurry that results from centrifugation will be that of either a dry mass (requiring resuspension but substantially free of residual broth, i.e., from a tubular bowl centrifuge) or a wet slurry (containing measurable residual broth and requiring additional resuspension). During the separation, conditions that result in cell lysis (such as extremes in temperature) must be avoided. In addition, while soluble protein is generally protected from shear and external proteolysis, these proteins are still subject to thermal denaturation.

Cell Disruption Intracellular protein products are present as either soluble, folded proteins or inclusion bodies. Intracellular protein products are very common because *Escherichia coli* is a main workhorse for recombinant proteins. *E. coli* is a gram-negative bacterium that precipitates recombinant proteins in the form of inclusion bodies. Release of folded proteins must be carefully considered. Active proteins are subject to deactivation and denaturation and thus require the use of "gentle" conditions. In addition, due consideration must be given to the suspending medium; lysis buffers are often optimized to promote protein stability and protect the protein from proteolysis and deactivation. Inclusion bodies, in contrast, are protected by virtue of the protein agglomeration. More stressful conditions are typically employed for their release, which includes going to higher temperatures if necessary. For "native" proteins, gentler methods and temperature control are required.

The release of intracellular protein product is achieved through rupture of the cell walls, and release of the protein product to the surrounding medium, through either mechanical or nonmechanical means, or through chemical, physical, or enzymatic lysis [Engler, in Cooney and Humphrey (eds.), op. cit., pp. 305–324; Schutte and Kula, in Stephanopoulos (ed.), op. cit., pp. 505–526]. Mechanical methods use pressure, as in the Manton/APV-Gaulin/French Press, or the Microfluidizer, or mechanical grinding, as in ball mills, the latter being used typically for flocs and usually only for natural products. Nonmechanical means include use of desiccants or solvents, while cell lysis can also be achieved through physical means (osmotic shock, freeze/thaw cycles), chemical (detergents, chaotropes), or enzymatic (lysozyme, phages).

In a high-pressure homogenizer, a pressurized cell suspension is forced through a valve and undergoes a rapid pressure change from up to 50 MPa to the atmospheric pressure. This results in the instant rupture of cells. Product release, which generally follows first-order kinetics, occurs through impingement of the high-velocity cell suspension jet on the stationary surfaces, and possibly also by the high-shear forces generated during the acceleration of the liquid through the gap. While sufficiently high pressures can be attained using commercially available equipment to ensure good release in a single pass, the associated adiabatic temperature increases (~1.8°C/1000 psig) may cause unacceptable activity losses for heat-labile proteins. Further denaturation can occur on exposure to the lysis medium. Thus, multiple passes may be preferred, with rapid chilling of the processed cell suspension between passes. The number of passes and the heat removal ability should be carefully optimized. The efficiency of the process depends on the homogenizing pressure and the choice of the valve unit, for which there are many designs available. Materials of construction are important to minimize erosion of the valve, to provide surface resistance to aggressive cleaning agents and disinfectants, and to permit steam cleaning and sanitization.

The release of inclusion bodies, in contrast, may follow a different strategy. Since inclusion bodies are typically recovered by centrifugation, it is often advantageous to send the lysate through the homogenizer with multiple passes to decrease the particle size of the cell debris. Since the inclusion bodies are much denser than the cell debris, the debris, now much reduced in size, can be easily separated from the inclusion bodies by centrifugation at low speeds. The inclusion bodies may be resuspended and centrifuged multiple times (often in the presence of low concentrations of denaturants) to clean up these aggregates. Since the inclusion bodies are already denatured, temperature control is not as important as in the case of native proteins.

Another popular method for cell disruption is to use a bead mill. In a bead mill, a cell suspension is mixed with glass or metal beads and agitated by using a rotating agitator at high speed. Bead mills have a controllable residence time compared with high-pressure homogenization.

However, they are susceptible to channeling and also fracturing of the beads. Tough cells require multiple passes to achieve a desired yield.

Chemical lysis, or solubilization of the cell wall, is typically carried out by using detergents such as Triton X-100, or the chaotropes urea, and guanidine hydrochloride. This approach does have the disadvantage that it can lead to some denaturation or degradation of the product. While favored for laboratory cell disruption, these methods are not typically used at the larger scales. Enzymatic destruction of the cell walls is also possible, and as more economical routes to the development of appropriate enzymes are developed, this approach could find industrial application. Again, the removal of these additives is an issue.

Physical methods such as osmotic shock, in which the cells are exposed to high salt concentrations to generate an osmotic pressure difference across the membrane, can lead to cell wall disruption. Similar disruption can be obtained by subjecting the cells to freeze/thaw cycles, or by pressurizing the cells with an inert gas (e.g., nitrogen) followed by a rapid depressurization. These methods are not typically used for large-scale operations.

On homogenization, the lysate may drastically increase in viscosity due to DNA release. This can be ameliorated to some extent by using multiple passes to reduce the viscosity. Alternatively, precipitants or nucleic acid digesting enzymes can be used to remove these viscosity-enhancing contaminants.

For postlysis processing, careful optimization must be carried out with respect to pH and ionic strength. Often it is necessary to do a buffer exchange. Cell debris can act as an ion exchanger and bind proteins ionically, thus not allowing them to pass through a filtration device or causing them to be spun out in a centrifuge. Once optimal conditions are found, these conditions can be incorporated in the lysis buffer by either direct addition (if starting from cell paste) or diafiltration (if starting from a cell concentrate).

Protein Refolding Although protein refolding is not a bioseparation operation, it is an integral part of a downstream process for the production of an inactive (typically intracellular) protein. So far, it still remains a challenging art casting an uncertainty on the success of a process. Early in the product development stage, animal cell culture may be required to produce many bioactive candidates for initial screening of efficacy. To commercialize the product protein, it may be reexpressed in a far more economic and productive host such as *E. coli.* The commercial products of recombinant DNA technology are frequently not produced in their native, biologically active form, because the foreign hosts such as *E. coli* in which they are produced lack the appropriate apparatus for the folding of the proteins. Thus, the overproduced proteins are generally recovered as refractile or inclusion bodies, or aggregates, typically 1 to 3 μm in size, and all cysteine residues are fully reduced. It is necessary at some stage in the processing to dissolve the aggregates and then refold them to obtain the desired biologically active product [Cleland and Wang, in Stephanopoulos (ed.), op. cit., pp. 527–555].

Advantages of inclusion bodies in the production stage are their ease of separation by centrifugation following cell disruption, because of their size and density, and their provision of excellent initial-purification possibilities, as long as impurities are not copurified to any significant extent with the inclusion bodies. They also provide a high expression level and prevent endogenous proteolysis. There can be, however, significant product loss during protein refolding to the active form. Figure 20-87 shows a typical process for refolding. The inclusion body which is released from the host cell by cell disruption is washed and then solubilized by using a denaturant such as guanidine hydrochloride (4 to 9 *M*), urea (7 to 8 *M*), sodium thiocyanate (4 to 9 *M*), or detergents such as Triton X-100 or sodium dodecyl sulfate. This step disrupts the hydrogen and ionic bonds to obtain fully denatured and stretched peptide chains. For proteins with disulfide bonds, addition of appropriate reducing agents (e.g., beta mercaptoethanol) is required to break all incorrectly formed intramolecular disulfide bonds. To permit proper refolding of the protein, it is necessary to remove the denaturant or detergent molecules from the surroundings of the stretched and solubilized peptides. This will initiate self-refolding of the protein molecules. For proteins with disulfide bonds, an oxidative reaction with oxygen or other oxidants is required to join two free SH groups to form an S—S covalent bond. This in vitro refolding operation is traditionally achieved by dilution with a refolding buffer. However, misfolding or aggregation is usually found in the refolding process. Analysis of in vitro refolding kinetics shows that there is at least an intermediate (I) between the unfolded protein (U) and the fully active refolded native protein (N), as illustrated below [Kuwajima and Arai, in R. H. Pain (ed.), *Mechanism of Protein Folding*, 2d ed., Oxford University Press, Oxford, 2000, pp. 138–171; Tsumoto et al., *Protein Expr. Purif.*, **28,** 1–8 (2003)].

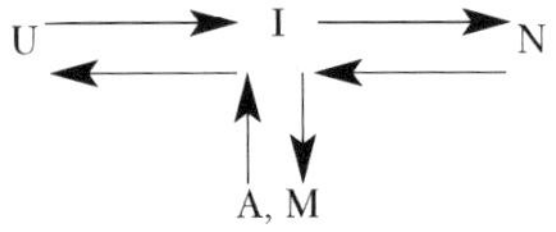

The process to convert U to I may be fast. The process for I to N, however, may be slow and highly reversible. Some intermediate molecules may form aggregates (A) or misfolded proteins (M). Figure 20-87 is a simplified refolding pathway. In reality, the situation can be more complicated where several intermediates (I_1, I_2, I_3, etc.) are present with numerous possibilities of aggregation and misfolding. For most proteins, refolding is a self-assembly process that follows a first-order kinetics. Aggregation, on the other hand, involves interactions between two or more molecules and follows the second- or higher-order kinetics. Therefore, in vitro refolding at higher protein concentrations would lead to the formation of more aggregates. Many observations have shown that a low final protein concentration, usually 10 to 100 μg/mL, is required for dilution refolding [Schlegl et al., *Chem. Engng. Sci.,* **60,** 5770–5780 (2005)]. In an industrial process, this strategy generally features large volumes of buffers, exerting an extra burden for subsequent purification steps because the concentration is low. Optimization of dilution strategy, such as the way of dilution, the speed of dilution, and the solution composition of a refolding buffer, is beneficial for an increased refolding yield.

A number of studies [Kuwajima and Arai, loc. cit.; Sadana, *Biotech. Bioeng.*, **48,** 481–489 (1995)] demonstrated that the presence of some molecules in the refolding buffer may suppress misfolding or aggregation. Molecular chaperones such as GroES and GroEL can promote correct refolding, but they are very expensive. Other molecules have been tried as artificial chaperones. The most commonly used molecules are polyethylene glycol (PEG) of various molecular weights and

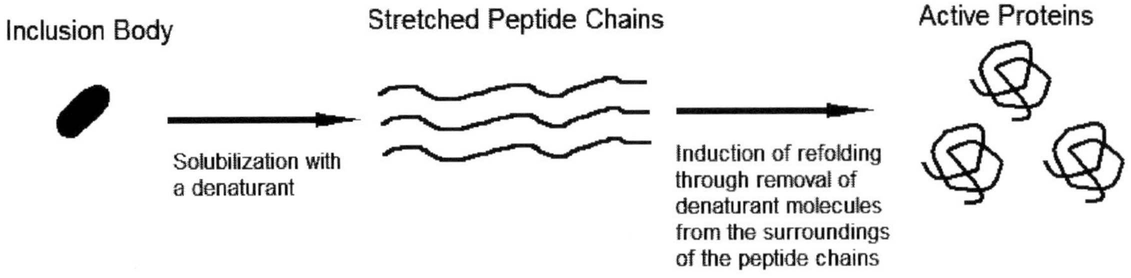

FIG. 20-87 Illustration of a refolding process for a protein from inclusion body.

concentrations, L-arginine (0.4 to 1 M), low concentrations of denaturants such as urea (1 to 2 M) and guanidine hydrochloride (0.5 to 1.5 M), and some detergents such as SDS, CTAB, and Triton X-100. The effects of these additives on protein refolding are still under investigation. A new direction of protein refolding involves the use of chromatographic techniques [Li et al., *Protein Expr. Purif.*, **33,** 1–10 (2004)]. Size exclusion, ion exchange, hydrophobic interaction, and metal chelate affinity chromatographic techniques have been studied with successful results. Chromatographic refolding explores the interaction between the protein molecule to be refolded and the packing medium in the column. It may reduce the interaction between protein molecules and increase the chance of self-assembly with the aid of the functional groups and pores in the matrix of the packing medium. An advantage for chromatographic refolding is the availability of gradient elution that creates a gradual change of the solution environment, leading to a gentle removal of the denaturant and a gradual change of favorable conditions such as pH and the artificial chaperone concentration. Simultaneous refolding and partial purification are possible with this new technique.

For extracellular products, which are invariably water-soluble, the first step is the removal of cells and cell debris using a clarification method and, in the case of typical protein products, the removal of dissolved low-molecular-weight compounds. This must be done under relatively gentle conditions to avoid undesired denaturation of the product. Again, either filtration or centrifugation can be applied. Filtration results in a cell-free supernatant with dilution associated with the diafiltration of the final cell slurry. Centrifugation, regardless of the mode, will result in a small amount of cells in the centrate, but there is no dilution of the supernatant. During the process development careful studies should be conducted to examine the effects of pH and ionic strength on the yield, as cells and cell debris may retain the product through charge interactions. If the broth or cell morphology does not allow for filtration or if dry cell mass is required, tubular-bowl centrifugation is typically utilized. Note that plant and animal cells cannot sustain the same degree of applied shear as can microbial cells, and thus crossflow filtration or classical centrifugation may not be applicable. Alternatives using low-shear equipment under gentle conditions are often employed in these situations.

For whole broths the range of densities and viscosities encountered affects the concentration factor that can be attained in the process, and can also render crossflow filtration uneconomical because of the high pumping costs, and so on. Often, the separation characteristics of the broth can be improved by broth conditioning using physicochemical or biological techniques, usually of a proprietary nature. The important characteristics of the broth are rheology and conditioning.

Clarification Using Centrifugation Centrifugation relies on the enhanced sedimentation of particles of density different from that of the surrounding medium when subjected to a centrifugal force field [Axelsson, in Cooney and Humphrey (eds.), op. cit., pp. 325–346] (see also Sec. 18, "Liquid-Solid Operations and Equipment"). Advantages of centrifugal separations are that they can be carried out continuously and have short retention times, from a fraction of a second to seconds, which limit the exposure time of sensitive biologicals to shear stresses. Yields are high, provided that temperature and other process conditions are adequately controlled. They have small space requirements, and an adjustable separation efficiency makes them a versatile unit operation. They can be completely closed to avoid contamination, and, in contrast to filtration, no chemical external aids are required that can contaminate the final product. The ability now to contain the aerosols typically generated by centrifuges adds to their operability and safety.

Sedimentation rates must be sufficiently high to permit separation, and they can be enhanced by modifying solution conditions to promote the aggregation of proteins or impurities. An increase in precipitation of the contaminating species can often be accomplished by a reduction in pH or an elevation in temperature. Flocculating agents, which include polyelectrolytes, polyvalent cations, and inorganic salts, can cause a 2000-fold increase in sedimentation rates. Some examples are polyethylene imine, EDTA, and calcium salts. Cationic bioprocessing aids (cellulosic or polymeric) reduce pyrogen, nucleic acid, and acidic protein loads which can foul chromatography columns. The removal of these additives during both centrifugation and subsequent processing must be clearly demonstrated.

There are many different types of centrifuges, classified according to the way in which the transport of the sediment is handled [Medronho, in Hatti-Kaul and Mattiasson (eds.), op. cit., pp. 131–190]. The selection of a particular centrifuge type is determined by its capacity for handling sludge; the advantages and disadvantages of various separator types are discussed by Axelsson [in Cooney and Humphrey (eds.), op. cit., pp. 325–346]. Solids-retaining centrifuges are operated in a semibatch mode, as they must be shut down periodically to remove the accumulated solids; they are primarily used when solids concentrations are low, and they have found application during the clarification and simultaneous separation of two liquids. In solids-ejecting centrifuges, the solids are removed intermittently either through radial slots or axially while the machine is running at full speed. These versatile machines can be used to handle a variety of feeds, including yeast, bacteria, mycelia, antibiotics, enzymes, and so on. Solids-discharging nozzle centrifuges have a large capacity and can accommodate up to 30 percent solids loading. Decanter centrifuges consist of a drum, partly cylindrical and partly conical, and an internal screw conveyor for transport of the solids, which are discharged at the conical end; liquids are discharged at the cylindrical end. Levels within the drum are set by means of external nozzles.

Continuous-flow units, the scroll decanter and disk-stack centrifuges, are easiest to use from an operational perspective; shutdown of the centrifuge during the processing of a batch is not expected. While the disk-stack centrifuge enjoys popularity as a process instrument within the pharmaceutical and biotechnology industries, the precise timing of solids ejection and the continuous high-speed nature of the device make for complex equipment and frequent maintenance. It is often used to harvest cells, since the solids generated are substantially wet and could lead to measurable yield losses in extracellular product systems. For intracellular product processing, the wet cell sludge is easily resuspended for use in subsequent processing.

The tubular bowl, in contrast, is a semibatch processing unit owing to the limited solids capacity of the bowl. The use of this unit requires shutdown of the centrifuge during the processing of the batch. The semibatch nature of these centrifuges can thus greatly increase processing cycle times. The introduction of disposable sheets to act as bowl liners has significantly impacted turnaround times during processing. The dry nature of the solids generated makes the tubular-bowl centrifuge well suited for extracellular protein processing, since losses to the cell sludge are minimal. In contrast, the dry, compact nature of the sludge can make the cells difficult to resuspend. This can be problematic for intracellular protein processing where cells are homogenized in easily clogged, mechanical disrupters.

Clarification Using Microfiltration Crossflow filtration (microfiltration includes crossflow filtration as one mode of operation in "Membrane Separation Processes," which appears earlier in this section) relies on the retention of particles by a membrane. The driving force for separation is pressure across a semipermeable membrane, while a tangential flow of the feed stream parallel to the membrane surface inhibits solids settling on and within the membrane matrix (Datar and Rosen, loc. cit.).

Microfiltration is used for the removal of suspended particles, recovery of cells from fermentation broth, and clarification of homogenates containing cell debris. Particles removed by microfiltration typically average greater than 500,000 nominal molecular weight [Tutunjian, in Cooney and Humphrey (eds.), op. cit., pp. 367–381; Gobler, in Cooney and Humphrey (eds.), op. cit., pp. 351–366]. Ultrafiltration focuses on the removal of low-molecular-weight solutes and proteins of various sizes, and it operates in the less than 100,000 nominal-molecular-weight cutoff (NMWCO) range [Le and Howell, in Cooney and Humphrey (eds.), op. cit., pp. 383–409]. Both operations consist of a concentration segment (of the larger particles) followed by diafiltration of the retentate [Tutunjian, in Cooney and Humphrey (eds.), op. cit., pp. 411–437].

Generally, the effectiveness of the separation is determined not by the membrane itself, but rather by the formation of a secondary or dynamic membrane caused by interactions of the solutes and particles with the membrane. The buildup of a gel layer on the surface of an ultrafiltration membrane owing to rejection of macromolecules can provide the primary separation characteristics of the membrane. Similarly, with colloidal suspensions, pore blocking and bridging of pore entries can modify the membrane performance, while molecules of size similar to the membrane pores can adsorb on the pore walls, thereby restricting passage of water and smaller solutes. Media containing poorly defined ingredients may contain suspended solids, colloidal particles, and gel-like materials that prevent effective microfiltration. In contrast to centrifugation, specific interactions can play a significant role in membrane separation processes.

The factors to consider in the selection of crossflow filtration include the flow configuration, tangential linear velocity, transmembrane pressure drop (driving force), separation characteristics of the membrane (permeability and pore size), size of particulates relative to the membrane pore dimensions, low protein-binding ability, and hydrodynamic conditions within the flow module. Again, since particle-particle and particle-membrane interactions are key, broth conditioning (ionic strength, pH, etc.) may be necessary to optimize performance.

Selection of Cell-Separation Unit Operation The unit operation selected for cell separations can depend on the subsequent separation steps in the train. In particular, when the operation following cell separation requires cell-free feed (e.g., chromatography), filtration is used, since centrifugation is not absolute in terms of cell separation. In addition, if cells are to be stored (i.e., they contain the desired product) because later processing is more convenient (e.g., only two-shift operation, facility competes for equipment with other products, batch is too big for single pass in equipment), it is generally better to store the cells as a frozen concentrate than a paste, since the concentrate thaws more completely, avoiding small granules of unfrozen cell solids that can foul homogenizers, columns, and filters. Here the retentate from filtration is desired, although the wet cell mass from a disc stack-type centrifuge may be used.

Centrifugation is generally necessary for complex media used to make natural products, for while the media components may be sifted prior to use, they can still contain small solids that can easily foul filters. The medium to be used should be tested on a filter first to determine the fouling potential. Some types of organisms, such as filamentous organisms, may sediment too slowly owing to their larger cross sections, and they are better treated by filtration (mycelia have the potential to easily foul tangential-flow units; vacuum-drum filtration using a filter aid, e.g., diatomaceous earth, should also be considered). Often the separation characteristics of the broth can be improved by broth conditioning using physicochemical or biological techniques, usually of a proprietary nature.

Regardless of the machine device, centrifuges are typically maintenance-intensive. Filters can be cheaper in terms of capital and maintenance costs and should be considered first unless centrifugal equipment already exists. Small facilities (< 1000 L) use filtration, since centrifugation scale-down is constrained by equipment availability. Comparative economics of the two classes of operations are discussed by Datar and Rosen (loc. cit.).

INITIAL PURIFICATION

Initial purification is the rough purification (considered by many people as isolation) to prepare a feed for subsequent high-resolution steps. In initial purification steps the goal is to obtain concentration with partial purification of the product, which is recovered as a precipitate (precipitation), a solution in a second phase (liquid-liquid partitioning), or adsorbed to solids (adsorption, chromatography).

Precipitation Precipitation of products, impurities, or contaminants can be induced by the addition of solvents, salts, or polymers to the solution; by increasing temperature; or by adjusting the solution pH (Scopes, op. cit., pp. 41–71; Ersson et al., in Janson and Ryden, op. cit., pp. 3–32). This operation is used most often in the early stages of the separation sequence, particularly following centrifugation, filtration, and/or homogenization steps. Precipitation is often carried out in two stages, the first to remove bulk impurities and the second to precipitate and concentrate the target protein. Generally, amorphous precipitates are formed, owing to occlusion of salts or solvents, or to the presence of impurities.

Salts can be used to precipitate proteins by "salting out" effects. The effectiveness of various salts is determined by the Hofmeister series, with anions being effective in the order citrate > PO_4^{2-} > SO_4^{2-} > CH_3COO^- > Cl^- > NO_3^-, and cations according to NH_4^+ > K^+ > Na^+ (Ersson et al., op. cit., p. 10; Belter et al., op. cit., pp. 221–236).

Salts should be inexpensive owing to the large quantities used in precipitation operations. Ammonium sulfate appears to be the most popular precipitant because it has an effective cation and an effective anion, high solubility, easy disposal, and low cost. Drawbacks to this approach include low selectivity, high sensitivity to operating conditions, and downstream complications associated with salt removal and disposal of the high-nitrogen-content stream. Generally, aggregates formed on precipitation with ammonium sulfate are fragile, and are easily disrupted by shear. Thus, these precipitation operations are, following addition of salt, often aged without stirring before being fed to a centrifuge by gravity feed or using low-shear pumps (e.g., diaphragm pumps).

The organic solvents most commonly used for protein precipitation are acetone and ethanol (Ersson et al., op. cit.). These solvents can cause some denaturation of the protein product. Temperatures below 0°C can be used, since the organic solvents depress the freezing point of the water. The precipitate formed is often an extremely fine powder that is difficult to centrifuge and handle. With organic solvents, in-line mixers are preferred, as they minimize solvent-concentration gradients and regions of high-solvent concentrations, which can lead to significant denaturation and local precipitation of undesired components typically left in the mother liquors. In general, precipitation with organic solvents at lower temperature increases yield and reduces denaturation. It is best carried out at ionic strengths of 0.05 to 0.2 *M*.

Water-soluble polymers and polyelectrolytes (e.g., polyethylene glycol, polyethylene imine polyacrylic acid) have been used successfully in protein precipitations, and there has been some success in affinity precipitations wherein appropriate ligands attached to polymers can couple with the target proteins to enhance their aggregation. Protein precipitation can also be achieved by using pH adjustment, since proteins generally exhibit their lowest solubility at their isoelectric point. Temperature variations at constant salt concentration allow for fractional precipitation of proteins.

Precipitation is typically carried out in standard cylindrical tanks with low-shear impellers. If in-line mixing of the precipitating agent is to be used, this mixing is employed just prior to the material's entering the aging tank. Owing to their typically poor filterability, precipitates are normally collected by using a centrifugal device.

Liquid-Liquid Partitioning Liquid-liquid partitioning (see also Sec. 15 on liquid-liquid extraction) involving an organic solvent is commonly known as solvent extraction or extraction. Solvent extraction is routinely used to separate small biomolecules such as antibiotics and amino acids. However, it is typically not suitable for protein fractionation with only a few exceptions because organic solvents may cause protein denaturation or degradation. A recent review of solvent extraction for bioseparations including a discussion on various parameters that can be controlled for solvent extraction was given by Gu [in Ahuja (ed.), op. cit., pp. 365–378]. As a replacement of solvent extraction, aqueous two-phase partitioning is typically used for protein purification. It uses two water-soluble polymers (and sometimes with some salts when polyelectrolytes are involved) to form two aqueous phases [Albertsson, op. cit.; Kula, in Cooney and Humphrey (eds.), op. cit., pp. 451–471]. Both phases contain water but differ in polymer (and salt) concentration(s). The high water content, typically greater than 75 percent, results in a biocompatible environment not attainable with traditional solvent extraction systems. Biomolecules such as proteins have different solubilities in the two phases, and this provides a basis for separation. Zaslavsky (op. cit., pp. 503–667) listed 163 aqueous two-phase systems including PEG-dextran-water, PEG-polyvinylmethylether-water, PEG-salt-water, polyvinylpyrrolidone-dextran-water, polyvinylalcohol-dextran-water, and Ficoll-dextran-water systems. Partitioning between the two aqueous phases is controlled by the polymer molecular weight and concentration, protein net charge and size, and hydrophobic and electrostatic interactions. Aqueous two-phase polymer systems are suitable for unclarified broths since particles tend to collect at the interface between the two phases, making their removal very efficient. They can also be used early on in the processing train for initial bulk-volume reduction and partial purification. One of the drawbacks of these systems is the subsequent need for the removal of phase-forming reagents.

Affinity partitioning is carried out by adding affinity ligands to an aqueous two-phase partitioning system. The biospecific binding of a biomolecule with the ligand moves the biomolecule to a preferred phase that enhances the partitioning of the biomolecule [Johansson and Tjerneld, in Street (ed.), *Highly Selective Separations in Biotechnology*, Blackie Academic & Professional, London, 1994, pp. 55–85]. Diamond and Hsu [in Fiechter (ed.), *Advances in Biochemical Engineering/Biotechnology,* vol. 47, Springer-Verlag, Berlin-New York, 2002, pp. 89–135] listed several dozens of biomolecules, many of which are proteins that have been separated by using affinity partitioning. Fatty acids and triazine are the two common types of affinity ligand while metallated iminodiacetic acid (IDA) derivatives of PEG such as Cu(II)IDA-PEG can be used for binding with proteins rich in surface histidines. The drawbacks of affinity partitioning include the costs of ligands and the need to couple the ligands to the polymers used in the aqueous two-phase partitioning.

Product recovery from these systems can be accomplished by changes in either temperature or system composition. Composition changes can be affected by dilution, backextraction, and micro- and ultrafiltration. As the value of the product decreases, recovery of the polymer may take on added significance. A flow diagram showing one possible configuration for the extraction and product and polymer recovery operations is shown in Fig. 20-88 [Greve and Kula, *J. Chem. Tech. Biotechnol.*, **50,** 27–42 (1991)]. The phase-forming polymer and salt are added directly to the fermentation broth. The cells or cell debris and contaminating proteins report to the salt-rich phase and are discarded. Following pH adjustment of the polymer-rich phase, more salt is added to induce formation of a new two-phase system in which the product is recovered in the salt phase, and the polymer can be recycled. In this example, disk-stack centrifuges are used to enhance the phase separation rates. Other polymer recycling options include extraction with a solvent or supercritical fluid, precipitation, or diafiltration. Electrodialysis can be used for salt recovery and recycling.

Reversed micellar solutions can also be used for the selective extraction of proteins [Kelley and Hatton, in Stephanopoulos (ed.), op. cit., pp. 593–616]. In these systems, detergents soluble in an oil phase aggregate to stabilize small water droplets having dimensions similar to those of the proteins to be separated. These droplets can host hydrophilic species such as proteins in an otherwise inhospitable organic solvent, thus enabling these organic phases to be used as protein extractants. Factors affecting the solubilization effectiveness of the solvents include charge effects, such as the net charge determined by the pH relative to the protein isoelectric point; charge distribution and asymmetry on the protein surface; and the type (anionic or cationic) of the surfactant used in the reversed micellar phase. Ionic strength and salt type affect the electrostatic interactions between the proteins and the surfactants, and affect the sizes of the reversed micelles. Attachment of affinity ligands to the surfactants has been demonstrated to lead to enhancements in extraction efficiency and selectivity [Kelley et al., *Biotech. Bioeng.*, **42,** 1199–1208 (1993)].

Product recovery from reversed micellar solutions can often be attained by simple backextraction, by contacting with an aqueous solution having salt concentration and pH that disfavors protein solubilization, but this is not always a reliable method. Addition of cosolvents such as ethyl acetate or alcohols can lead to a disruption of the micelles and expulsion of the protein species, but this may also lead to protein denaturation. These additives must be removed by distillation, e.g., to enable reconstitution of the micellar phase. Temperature increases can similarly lead to product release as a concentrated aqueous solution. Removal of the water from the reversed micelles by molecular sieves or silica gel has also been found to cause a precipitation of the protein from the organic phase.

Extraction using liquid emulsion membranes involves the use of a surface-active agent such as a surfactant to form dispersed droplets that encapsulate biomolecules. Its economic viability for large-scale applications is still weak [Patnaik, in Subramanian (ed.), op. cit., vol. 1, pp. 267–303]. Another less-known method involving the use of a surfactant is the foam fractionation method that has seen limited applications. Ionic fluids have found commercial applications in chemical reactions by replacing volatile solvents. They are emerging as an environmentally friendly solvent replacement in liquid-liquid phase partitioning. Room-temperature ionic liquids are low-melting-point salts that stay as liquids at room temperature. Partition behavior in a system involving a room-temperature ionic fluid and an aqueous phase is influenced by the type of ionic liquid used as well as pH change (Visser et al., *Green Chem.*, Feb. 1–4, 2000). An ionic fluid was also reportedly used in the mobile phase for liquid chromatography [He et al., *J. Chromatogr. A*, **1007,** 39–45 (2003)]. More research needs to be done in this area to develop this new green technology [Freemantle, *C&EN*, **83,** 33–38 (2005).]

Aqueous-detergent solutions of appropriate concentration and temperature can phase-separate to form two phases, one rich in detergents, possibly in the form of micelles, and the other depleted of the detergent [Pryde and Phillips, *Biochem. J.*, **233,** 525–533 (1986)].

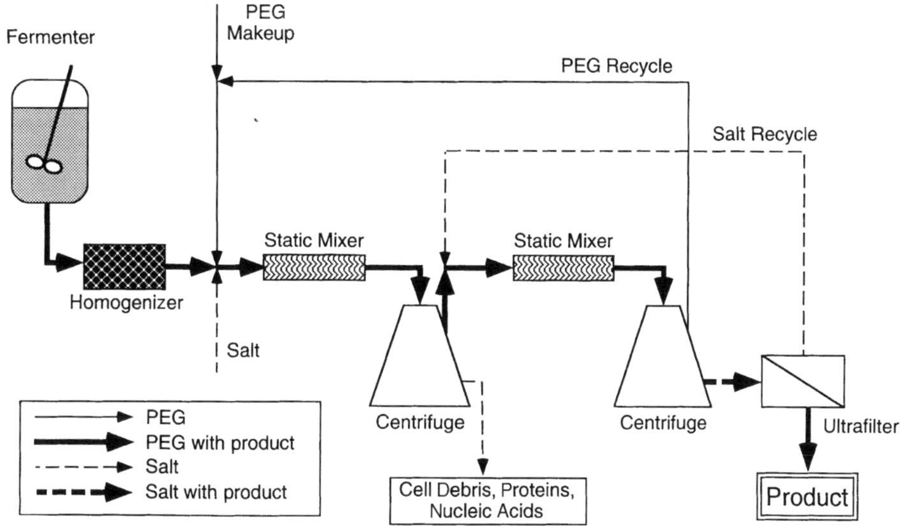

FIG. 20-88 Process scheme for protein extraction in aqueous two-phase systems for the downstream processing of intracellular proteins, incorporating PEG and salt recycling. [*Reprinted from Kelly and Hatton in Stephanopoulos (ed.), op. cit.; adapted from Greve and Kula, op. cit.*]

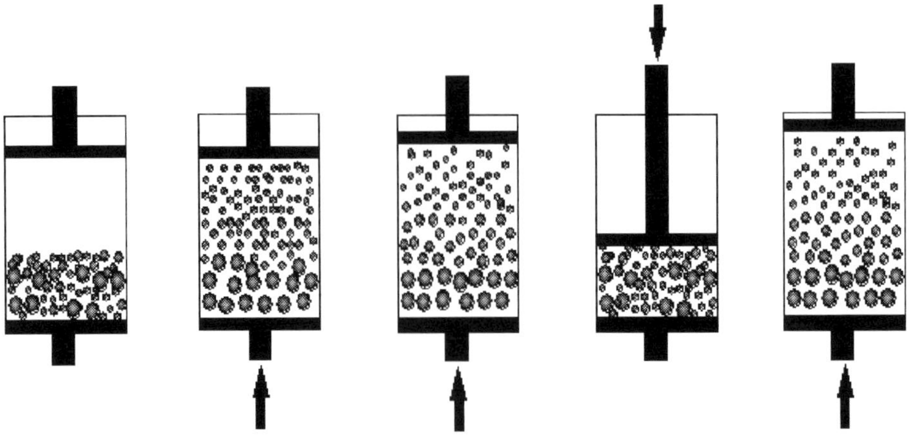

FIG. 20-89 An operation cycle in expanded-bed adsorption.

Proteins distribute between the two phases, hydrophobic (e.g., membrane) proteins reporting to the detergent-rich phase and hydrophilic proteins to the detergent-free phase. Indications are that the size-exclusion properties of these systems can also be exploited for viral separations. These systems would be handled in the same way as the aqueous two-phase systems.

On occasion, for extracellular products, cell separation can be combined with an initial volume reduction and purification step by using liquid-liquid extraction. This is particularly true for low-molecular-weight products, and it has been used effectively for antibiotic and vitamin recovery. Often scroll decanters can be used for this separation. The solids are generally kept in suspension (which requires that the solids be denser than heavy phase), while the organic phase, which must be lighter than water (cells typically sink in water), is removed. Experience shows that scrolls are good for handling the variability seen in fermentation feedstock. Podbielniak rotating-drum extraction units have been used often, but only when solids are not sticky, gummy, or flocculated, as they can get stuck in perforations of the concentric drums, but will actually give stages to the extraction in short-residence time (temperature-sensitive product). The Karr reciprocating-plate column can handle large volumes of whole-broth materials efficiently, and it is amenable to ready scale-up from small laboratory-scale systems to large plant-scale equipment.

Adsorption Adsorption (see also Sec. 16, "Adsorption and Ion Exchange") can be used for the removal of pigments and nucleic acids, e.g., or can be used for direct adsorption of the desired proteins. Stirred-batch or expanded-bed operations allow for presence of particulate matter, but fixed beds are not recommended for unclarified broths owing to fouling problems. These separations can be effected through charge, hydrophobic, or affinity interactions between the species and the adsorbent particles, as in the chromatographic steps outlined below. The adsorption processes described here are different from those traditionally ascribed to chromatography in that they do not rely on packed-bed operations.

In continuous affinity recycle extraction (CARE) operations, the adsorbent beads are added directly to the cell homogenate, and the mixture is fed to a microfiltration unit. The beads loaded with the desired solute are retained by the membrane, and the product is recovered in a second stage by changing the buffer conditions to disfavor binding.

Expanded-bed adsorption (EBA) has gained popularity in bioprocessing since its commercial introduction in the 1990s because of its ability to handle a crude feedstock that contains cells or other particulates. EBA eliminates the need for a dedicated clarification step by combining solid-liquid separation and adsorption into a single-unit operation [Hjorth et al., in Subramanian (ed.), op. cit., vol. 1, pp. 199–226; Mattiasson et al., in Ahuja (ed.), op. cit., pp. 417–451]. A typical EBA operation cycle is illustrated in Fig. 20-89. A bed packed with an adsorption medium (or a resin), usually spherical particles of different sizes, is expanded by an upward-flow liquid stream from the bottom. An unclarified feed is introduced after a stable expansion of the bed is achieved. Particulates pass through the void spaces between the resin particles, while the soluble product molecules are adsorbed by the resin and retained in the column. After a washing step, the resin particles are left to settle in the column to form a fixed bed. The product molecules are then eluted out with a mobile phase entering from the top of the column in a way similar to that in conventional fix-bed chromatography, to achieve a high-resolution separation. The elution can also be performed in expanded mode if needed. The regeneration step in the expanded mode flushes away residual particulates and refreshes the media for the next cycle. The difference between EBA and conventional fluidized-bed adsorption lies in the adsorption resin. In conventional fluidized-bed adsorption, the resin particles are randomly distributed in the column. In EBA, however, the resin particles are distributed vertically with large ones near the column bottom and the small ones near the top. There is no backmixing along the axial direction of the vertically standing column, thus achieving adsorption similar to that in a fixed-bed column. The resin particles have to be prepared to possess a suitable size distribution, or alternatively a distribution based on density differences if the particles sizes are uniform. An EBA column should have a length typically 3 to 4 times of the settled bed height to allow for bed expansion. An adjustable adapter at the top is needed to push the resin downward for elution in the fixed-bed mode. A proper design of the bottom frit is critical. Its holes must be smaller enough to retain the smallest resin particles but large enough to allow the particulates to enter and exit the column freely. In practical applications, plugging of the frit and nonuniform upward flow tend to be problematic, especially in columns with large diameters. To mitigate this problem, some EBA columns use an optimized inlet design and a mechanical stirrer at the bottom.

The advantages of EBA are its ability to adsorb the soluble product molecules directly from a cell suspension, a cell homogenate, or a crude biological fluid containing various particulates, thus making the "whole-broth processing" concept a reality. EBA eliminates the solid-liquid separation step (such as microfiltration and centrifugation) and enables a fast, more compact process requiring fewer steps and less time. It performs solid removal, concentration, and purification, all in a single-unit operation. By doing so, it can also minimize the risk of proteolytic breakdown of the product.

Membrane Ultrafiltration Membrane ultrafiltration is often one of the favored unit operations used for the isolation and concentration of biomolecules because they can be easily scaled up to process large feed volumes at low costs. Toward the end of an ultrafiltration operation, additional water or buffer is added to facilitate the passage of smaller molecules. This is known as diafiltration. Diafiltration is especially helpful in the removal of small contaminating species such as unspent nutrients including salts and metabolites. Salt removal is usually necessary if the next step is ion-exchange or reverse-phase chromatography. For a protein that is not very large, two ultrafiltration steps can be used in sequence. In the first one, the protein ends up in the permeate, allowing the removal of large contaminating molecules including pyrogens and also viral particles. In the second ultrafiltration, the protein stays in the retentate. This removes small contaminating molecules and concentrates the feed up to the protein's solubility limit. Membrane materials, configurations, and design considerations were discussed earlier in this section. Proper membrane materials must be selected to avoid undesirable binding with proteins. External fouling, pore blockage, and internal fouling were discussed by Ghosh (*Protein Bioseparation Using Ultrafiltration*, Imperial College Press, London, 2003).

FINAL PURIFICATION

The final purification steps are responsible for the removal of the last traces of impurities. The volume reduction in the earlier stages of the separation train is necessary to ensure that these high-resolution operations are not overloaded. Generally, chromatography is used in these final stages. Electrophoresis can also be used, but since it is rarely found in process-scale operations, it is not addressed here. The final product preparation may require removal of solvent and drying, or lyophilization of the product.

Chromatography Liquid chromatography steps are ubiquitous in the downstream processing. It is the most widely used downstream processing operation because of its versatility, high selectivity, and efficiency, in addition to its adequate scale-up potential based on wide experience in the biochemical processing industries. As familiarity is gained with other techniques such as liquid-liquid extraction, they will begin to find greater favor in the early stages of the separation train, but are unlikely to replace chromatography in the final stages, where high purities are needed.

Chromatography is typically a fixed-bed adsorption operation, in which a column filled with chromatographic packing materials is fed with the mixture of components to be separated. Apart from size-exclusion (also known as gel-permeation or gel-filtration) chromatography, in the most commonly practiced industrial processes the solutes are adsorbed strongly to the packing materials until the bed capacity has been reached. The column may then be washed to remove impurities in the interstitial regions of the bed prior to elution of the solutes. This latter step is accomplished by using buffers or solvents which weaken the binding interaction of the proteins with the packings, permitting their recovery in the mobile phase. To minimize product loss of a high-value product, a small load far below the saturation capacity is applied to the column. A complete baseline separation can then be achieved after elution. Gradient elution uses varied modifier strength in the mobile phase to achieve better separations of more chemical species.

Types of Chromatography Practiced Separation of proteins by using chromatography can exploit a range of different physical and chemical properties of the proteins and the chromatography adsorption media [Janson and Ryden, op. cit.; Scopes, op. cit.; Egerer, in Finn and Prave (eds.), *Biotechnology Focus 1*, Hanser Publishers, Munich, 1988, pp. 95–151]. Parameters that must be considered in the selection of a chromatographic method include composition of the feed, the chemical structure and stability of the components, the electric charge at a defined pH value and the isoelectric point of the proteins, the hydrophilicity and hydrophobicity of the components, and molecular size. The different types of interactions are illustrated schematically in Fig. 20-90.

Ion-exchange chromatography relies on the coulombic attraction between the ionized functional groups of proteins and oppositely charged functional groups on the chromatographic support. It is used to separate the product from contaminating species having different charge characteristics under well-defined eluting conditions, and for concentration of the product, owing to the high-adsorptive capacity of most ion-exchange resins and the resolution attainable. Elution is carried out by using a mobile phase with competing ions or varied pH. Ion-exchange chromatography is used effectively at the front end of a downstream processing train for early volume reduction and purification.

The differences in sizes and locations of hydrophobic pockets or patches on proteins can be exploited in *hydrophobic interaction chromatography (HIC)* and *reverse-phase chromatography (RPC)*; discrimination is based on interactions between the exposed hydrophobic residues and hydrophobic ligands which are distributed evenly throughout a hydrophilic porous matrix. As such, the binding characteristics complement those of other chromatographic methods, such as ion-exchange chromatography. In HIC, the hydrophobic interactions are relatively weak, often driven by salts in high concentration, and depend primarily on the exposed residues on or near the protein surface; preservation of the native, biologically active state of the protein is a desirable feature of HIC. HIC's popularity is on the rise in recent years because of this feature. Elution can be achieved differentially by decreasing salt concentration or increasing the concentration of polarity perturbants (e.g., ethylene glycol) in the eluent.

Reverse-phase chromatography relies on significantly stronger hydrophobic interactions than in HIC, which can result in unfolding and exposure of the interior hydrophobic residues, i.e., leads to protein denaturation and irreversible inactivation; as such, RPC depends on total hydrophobic residue content. Elution is effected by organic solvents applied under gradient conditions. RPC is the most commonly used analytical chromatographic method due to its ability to separate a vast array of chemicals with high resolutions. Denaturation of proteins does not influence the analytical outcome unless protein precipitation in the mobile phase becomes a problem.

HIC typically uses polymer-based resins with phenyl, butyl, or octyl ligands while RPC uses silica beads with straight-chain alkanes with 4, 8, or 18 carbons. Larger ligands provide stronger interactions. Polymeric beads are used in RPC when basic pH is involved because silica beads are unstable at such pH. In HIC, the mobile phase remains an aqueous salt solution, while RPC uses solvent in its mobile phase to regulate binding. Raising the temperature increases the hydrophobic interactions at the temperatures commonly encountered in biological processing.

HIC is most effective during the early stages of a purification strategy and has the advantage that sample pretreatment such as dialysis or desalting after salt precipitation is not usually required. It is also finding increased use as the last high-resolution step to replace gel filtration. It is a group separation method, and generally 50 percent or more of extraneous impurities are removed. This method is characterized by high adsorption capacity, good selectivity, and satisfactory yield of active material.

Despite the intrinsically nonspecific nature of ion-exchange and reversed-phase/hydrophobic interactions, it is often found that chromatographic techniques based on these interactions can exhibit remarkable resolution. This is attributed to the dynamics of multisite interactions being different for proteins having differing surface distributions of hydrophobic and/or ionizable groups.

Size-exclusion chromatography's (SEC's) separation mechanism is based on the sizes and shapes of proteins and impurities. The effective size of a protein is determined by its steric geometry and solvation characteristics. Smaller proteins are able to penetrate the small pores in the beads while large proteins are excluded, making the latter elute out of column more quickly. To suppress the ion-exchange side effects, a salt is typically added to the mobile phase. Ammonium carbonate or bicarbonate is used if the salt is to be removed by sublimation alone during lyophilization. In rare cases, a solvent at a low concentration is added to the mobile phase to suppress hydrophobic interactions between the protein molecules and the stationary phase. In industrial production processes, SEC columns are used to separate small molecules from proteins. It is also a choice for desalting and buffer exchange in the product polishing stage. SEC columns are typically very large because the feed loads to SEC columns are limited to 3 to 5 percent of the bed volumes. This loading capacity is far less than those with packings that have binding interactions.

Protein affinity chromatography can be used for the separation of an individual compound, or a group of structurally similar compounds from crude-reaction mixtures, fermentation broths, or cell lysates by exploiting very specific and well-defined molecular interactions

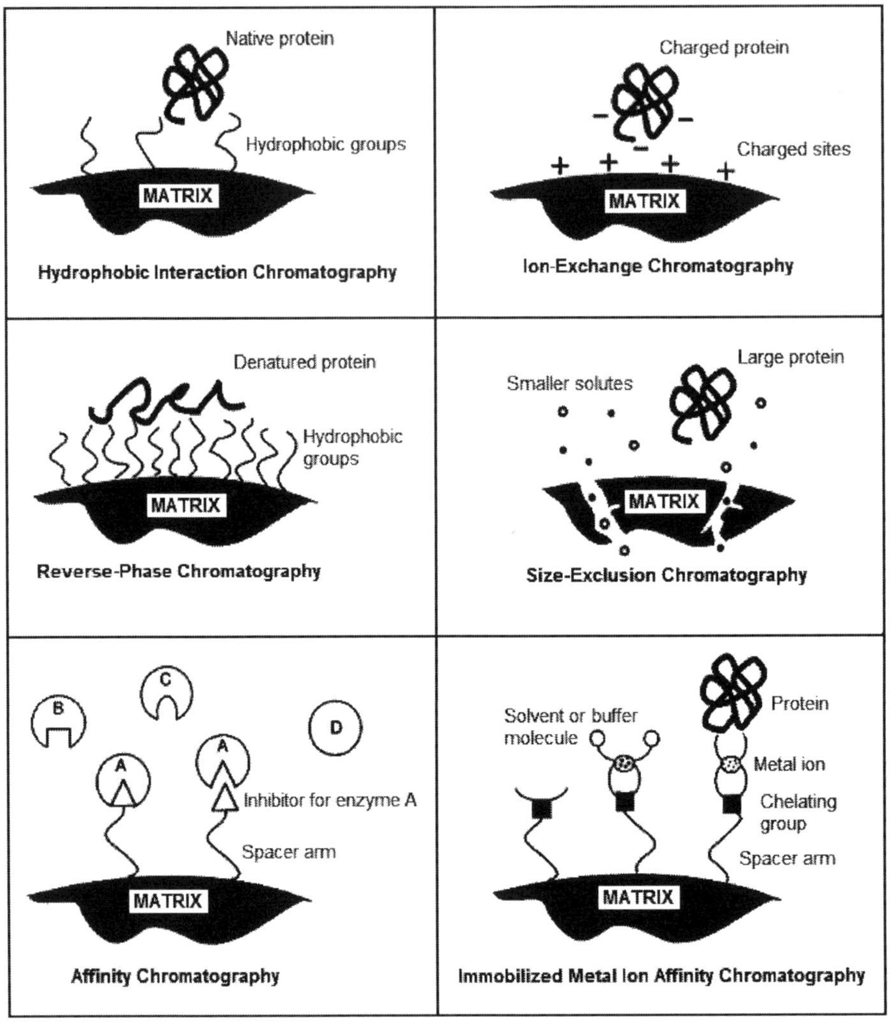

FIG. 20-90 Schematic illustration of the chromatographic methods most commonly used in downstream processing of protein products.

between the protein and affinity groups immobilized on the packing-support material. Examples of affinity interactions include antibody-antigen, hormone-receptor, enzyme-substrate/analog/inhibitor, metal ion–ligand, and dye-ligand pairs. Monoclonal antibodies are particularly effective as biospecific ligands for the purification of pharmaceutical proteins. Affinity chromatography may be used for the isolation of a pure product directly from crude fermentation mixtures in a single chromatographic step. Immunosorbents should not be subjected to crude extracts, however, as they are particularly susceptible to fouling and inactivation. Despite its high resolution and the ability to treat a very dilute feed, affinity chromatography is still costly on the process scale if protein ligands such as protein A or protein G is to be used, or a custom affinity matrix is required. Considerable research efforts are devoted to its development in part due to the increased number of protein pharmaceuticals produced at low concentrations.

After more than two decades of development, membrane chromatography has emerged as an attractive alternative to packed column chromatography. Using a porous membrane as the stationary phase in liquid chromatography has several potential advantages that include a very high flow rate through a very short and wide bed with only a modest transmembrane pressure drop. Elimination or minimization of diffusional mass-transfer resistance shifts the rate-controlling step to faster-binding kinetics, resulting in adsorptive separation of proteins in a fraction of the time required by conventional packed columns. To achieve sufficient adsorptive separation, it is necessary to use a medium that binds strongly with target molecules when a very short flow path is involved. Thus, membrane chromatography typically uses an affinity membrane, and the combination of membrane chromatography with affinity interaction provides high selectivity and fast processing for the purification of proteins from dilute feeds. To a much less extent, ion-exchange, hydrophobic interaction, and reverse-phase membrane chromatography have also been reported [Charcosset, *J. Chem. Technol. Biotechnol.* **71,** 95–110 (1998)].

Figure 20-91 shows the interaction between proteins in the mobile phase and the affinity membrane matrix. The mechanical strength, hydrophobicity, and ligand density of the membranes can be engineered through chemical modifications to make them suitable for affinity membrane chromatography. A preferred membrane medium to be prepared for affinity membrane chromatography should provide (1) desirable physical characteristics such as pore structure and mechanical strength, allowing fast liquid flow at a small pressure loss; (2) reactive groups (such as —OH, $—NH_2$, —SH, —COOH) for coupling ligands or spacer arms; (3) physical and chemical stabilities to endure heat or chemical sterilization; and (4) a nondenaturing matrix to retain protein bioactivity [Zou et al., *J. Biochem. Biophys. Methods,* **49,** 199–240 (2001)]. Base membranes

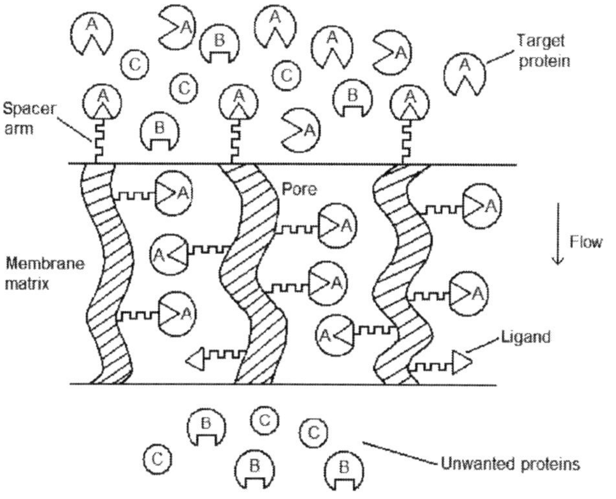

FIG. 20-91 Illustration of the interactions between proteins and membrane matrix in affinity membrane chromatography.

can be chosen from commercial membrane materials including organic, inorganic, polymeric, and composite materials. The selected membrane is first activated to create functional groups for chemical attachment of affinity ligands. If there is steric hindrance to binding between the immobilized ligand and the target molecule, a suitable spacer arm is used to bridge the activated membrane surface and the ligand. The ligand should retain its reversible binding capacity after immobilization onto the support membrane. Affinity ligands typically fall into two categories: (1) those derived from enzyme/substrate, antibody/antigen pairs that are capable of very strong and highly biospecific binding and (2) protein A and protein G, coenzyme, lectin, dyes, and metal chelates, etc., each capable of binding with a whole class of molecules. A highly specific ligand provides an unsurpassed resolution and an ability to handle a large volume of a dilute feed. However, they are typically fragile and expensive and may be unavailable off the shelf due to their narrow applications involving just one or a few molecules that can bind. Elution can also prove to be a difficult task because some biospecific bindings can be extremely tenacious. In contrast, a somewhat less specific ligand has a much wider market and thus is considerably less expensive. Various membrane cartridges have been used for affinity membrane chromatography including those with multiple layers of flat sheet membranes, hollow fibers, and spiral-wound and Chromarod membranes (Zou et al., loc. cit.).

Immobilized metal-ion affinity chromatography (IMAC) relies on the interaction of certain amino acid residues, particularly histidine, cysteine, and tryptophan, on the surface of the protein with metal ions fixed to the support by chelation with appropriate chelating compounds, invariably derivatives of iminodiacetic acid. Commonly used metal ions are Cu^{2+}, Zn^{2+}, Ni^{2+}, and Co^{2+}. Despite its relative complexity in terms of the number of factors that influence the process, IMAC is beginning to find industrial applications. The choice of chelating group, metal ion, pH, and buffer constituents will determine the adsorption and desorption characteristics. Elution can be effected by several methods, including pH gradient, competitive ligands, organic solvents, and chelating agents.

Following removal of unbound materials in the column by washing, the bound substances are recovered by changing conditions to favor desorption. A gradient or stepwise reduction in pH is often suitable. Otherwise, one can use competitive elution with a gradient of increasing concentration. IMAC eluting agents include ammonium chloride, glycine, histamine, histidine, or imidazol. Inclusion of a chelating agent such as EDTA in the eluent will allow all proteins to be eluted indiscriminately along with the metal ion.

Chromatographic Development The basic concepts of chromatographic separations are described elsewhere in this handbook. Proteins differ from small solutes in that the large number of charged and/or hydrophobic residues on the protein surface provide multiple binding sites, which ensure stronger binding of the proteins to the adsorbents, as well as some discrimination based on the surface distribution of amino acid residues. The proteins are recovered by elution with a buffer that reduces the strength of this binding and permits the proteins to be swept out of the column with the buffer solution. In isocratic elution, the buffer concentration is kept constant during the elution period. Since the different proteins may have significantly different adsorption isotherms, the recovery may not be complete, or it may take excessive processing time and cause excessive band spreading to recover all proteins from the column. In gradient elution operations, the composition of the mobile phase is changed during the process to decrease the binding strength of the proteins successively, the more loosely bound proteins being removed first before the eluent is strengthened to enable recovery of the more strongly adsorbed species. The change in eluent composition can be gradual and continuous, or it can be stepwise. Industrially, in large-scale columns it is difficult to maintain a continuous gradient owing to difficulties in fluid distribution, and thus stepwise changes are still used. In some adsorption modes, the protein can be recovered by the successive addition of competing compounds to displace the adsorbed proteins. In all cases, the product is eluted as a chromatographic peak, with some possible overlap between adjacent product peaks.

Displacement chromatography relies on a different mode of elution. Here a displacer that is more strongly adsorbed than any of the proteins is introduced with the mobile phase. As the displacer concentration front develops, it pushes the proteins ahead of itself. The more strongly adsorbed proteins then act as displacers for the less strongly bound proteins, and so on. This leads to the development of a displacer train in which the different molecules are eluted from the column in abutting roughly rectangular peaks in the reverse order of their binding strength with the column's stationary phase. Many displacers have been developed in the past two decades including both high- and low-molecular-weight molecules. Displacement chromatography may be practical as an earlier chromatography step in a downstream process when baseline separations are not necessary. It has been considered for some industrial processes. However, displacer reuse and possible contamination of the protein product in addition to its inherent inability to achieve a clear-cut baseline separation make the use of displacement chromatography still a challenge. So far, no FDA-approved process exists [Shukla and Cramer, in Ahuja (ed.), op. cit., pp. 379–415].

For efficient adsorption it is advisable to equilibrate both the column and the sample with the optimum buffer for binding. Prior to this, the column must be cleaned to remove tightly bound impurities by increasing the salt concentration beyond that used in the product elution stages. At the finish of cleaning operation, the column should be washed with several column volumes of the starting buffer to remove remaining adsorbed material. In desorption, it is necessary to drive the favored binding equilibrium for the adsorbed substance from the stationary to the mobile phase. Ligand-protein interactions are generally a combination of electrostatic, hydrophobic, and hydrogen bonds, and the relative importance of each of these and the degree of stability of the bound protein must be considered in selecting appropriate elution conditions; frequently compromises must be made. Gradient elution often gives good results.

Changes in pH or ionic strength are generally nonspecific in elution performance; ionic strength increases are effective when the protein binding is predominantly electrostatic, as in IEC. Polarity changes are effective when hydrophobic interactions play the primary role in protein binding. By reducing the polarity of the eluting mobile phase, this phase becomes a more thermodynamically favorable environment for the protein than adsorption to the packing support. A chaotropic salt (KSCN, KCNO, KI in range of 1 to 3 M) or denaturing agent (urea, guanidine HCl; 3 to 4 M) in the buffer can also lead to enhanced desorption. For the most hydrophobic proteins (e.g., membrane proteins) one can use detergents just below their critical micelle concentrations to solubilize the proteins and strip them from the packing surface. Specific elution requires more selective eluents. Proteins can be desorbed from ligands by competitive binding of the eluting agent (low concentration of 5 to 100 mM) either to the ligand or to the protein.

Specific eluents are most frequently used with group-specific adsorbents since selectivity is greatly increased in the elution step.

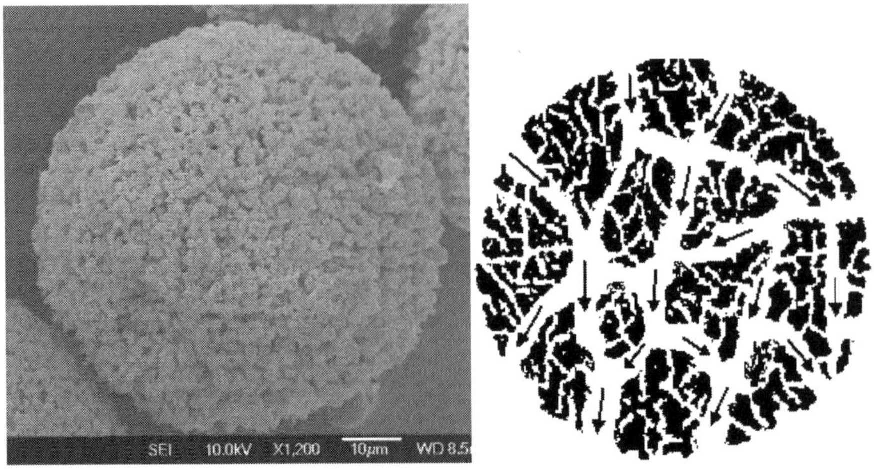

FIG. 20-92 SEM image of a poly(styrene-*co*-divinylbenzene) gigaporous particle synthesized from suspension polymerization and schematic of a gigaporous particle showing through-pores and diffusion pores [Gu et al., *China Particuology*, **3,** 349 (2005)].

The effectiveness of the elution step can be tailored by using a single eluent, pulses of different eluents, or eluent gradients. These systems are generally characterized by mild desorption conditions. If the eluting agent is bound to the protein, it can be dissociated by desalting on a gel filtration column or by diafiltration.

Column Packings The quality of the separation obtained in chromatographic separations will depend on the capacity, selectivity, and hydraulic properties of the stationary phase, which usually consists of porous beads of hydrophilic polymers filled with the solvent. The xerogels (e.g., crosslinked dextran) shrink and swell depending on solvent conditions, while aerogels have sizes independent of solution conditions. A range of materials are used for the manufacture of gel beads, classified according to whether they are inorganic, synthetic, or polysaccharides. The most widely used materials are based on neutral polysaccharides and polyacrylamide. Cellulose gels, such as crosslinked dextran, are generally used as gel filtration media, but can also be used as a matrix for ion exchangers. The primary use of these gels is for desalting and buffer exchange of protein solutions, as nowadays fractionation by gel filtration is performed largely with composite gel matrices. Agarose, a low-charge fraction of the seaweed polysaccharide agar, is a widely used packing material.

Microporous gels made by point crosslinking dextran or polyacrylamides are used for molecular-sieve separations such as size-exclusion chromatography and gel filtration, but are generally too soft at the porosities required for efficient protein chromatography. Macroporous gels are most often obtained from aggregated and physically crosslinked polymers. Examples include agarose, macroreticular polyacrylamide, silica, and synthetic polymers. These gels are good for ion-exchange and affinity chromatography as well as for other adsorption chromatography techniques. Composite gels, in which the microporous gel is introduced into the pores of macroreticular gels, combine the advantages of both types.

High matrix rigidity is offered by porous inorganic silica, which can be derivatized to enhance its compatibility with proteins, but it is unstable at alkaline pH. Hydroxyapatite particles have high selectivity for a wide range of proteins and nucleic acids.

Traditional porous media have pore sizes typically in the range of 100 to 300 Å. To reduce intraparticle diffusion for fast chromatography, column packing materials can be made either nonporous or extremely porous. Gustavsson and Larsson [in Hatti-Kaul and Mattiasson (eds), op. cit., pp. 423–454] discussed various chromatography media for fast chromatography of proteins. Nonporous particles such as the popular 5-μm modified silica beads for reverse-phase chromatography are excellent for analytical applications. However, due to their limited binding sites that exist only on the outer surface of the particles, nonporous particles are not suitable for preparative- or large-scale applications. Gigaporous media are gaining momentum in recent years. They are particles with pore sizes above 1000 Å. Some have large enough interconnecting pores in the range of 4000 to 8000 Å that allow even convective flow inside the particles. POROS® perfusion chromatography media are the first commercial products in this category introduced in the 1980s. A few more products are being commercialized. POROS® media are synthesized in two steps. Nanosize subparticles are first synthesized and then polymerized to form 10- to 50-μm particles in a second step. In recent years, advances in suspension polymerization produced a new type of integral gigaporous media with improved physical strength. The spherical particles are formed in a single polymerization step. An oil phase (dispersed phase) consisting of a monomer (such as styrene), a crosslinking agent (such as divinylbenzene), an initiator, a diluent, and a special porogen is used. By dispersing the oil phase in a water phase containing a stabilizer, a suspension can be obtained. The suspension polymerization is carried out at an elevated temperature above the decomposition temperature of the initiator to obtain the polymer particles. The diluent and porogen in the particles form smaller diffusion pores and much larger through-pores, respectively. Figure 20-92 shows such a particle with both conventional diffusion pores and gigaporous through-pores. The through-pores allow convective flow to improve mass transfer, and the diffusion pores provide an overall large surface area required for a large binding capacity.

In the scale-up of a chromatographic column with conventional packing media, the column length scale-up is limited because of the subsequently increased column pressure. To increase the feed load, the column is typically scaled up by enlarging the column diameter after the column length reaches a certain limit. The resultant pancake-shaped column leads to a deteriorating resolution due to poor flow distribution in the column cross-section. With rigid gigaporous particles, axial direction scale-up becomes possible because the flow rate can be up to 40 times higher than that for the conventional media. This kind of scale-up is far superior because of the enhanced resolution together with increased feed capacity when the column axial length is increased. With the advances made in rigid gigaporous media, chromatography columns can potentially process large volumes of very dilute feeds when combined with strong binding kinetics such as those involved in affinity chromatography. This may have a significant impact on the overall downstream process design.

Alternative Chromatographic Columns Commercial radial-flow chromatography (RFC) columns first appeared in the mid-1980s. RFC is an alternative to the conventional axial-flow chromatography for preparative- and large-scale applications. In a RFC column, the mobile phase flows in the radial direction rather than the axial direction. Computer simulation proves that RFC is somewhat equivalent to a pancakelike axial-flow column [Gu, in Flickinger and Drew (eds.), op. cit., pp. 627–639]. Both configurations offer a short flow path and thus a low pressure drop.

Continuous chromatography can be achieved by using a simulated moving-bed (SMB) process shown in Sec. 16, in which several columns are linked with a switching device to simulate a continuous countercurrent flow process. This design maximizes productivity while minimizing eluent consumption [Nicoud, in Ahuja (ed.), op. cit., pp. 475–509]. It is suitable for a simple feed that results in only a limited fraction [Imamoglu, in Freitag (ed.), op. cit., pp. 212–231]. By switching the feed, eluant input positions and raffinate, extract output positions periodically in a series of columns to simulated countercurrent movement of the liquid phase and the solid (resin) phase, continuous chromatography is obtained through the use of multiple columns in series. A four-zone SMB consisting of four columns is capable of producing two fractions as a pair of raffinate and exact. To have two pairs (with a total of four fractions) of raffinate and extract outputs an eight-zone SMB system is needed. Another alternative design described in Sec. 16 is the so-called annual-flow column that rotates continuously in the angular direction. Despite its obvious advantage of being straightforwardly continuous, it suffers from angular dispersion and reduced bed volume. This design so far has seen very limited application since its commercial introduction in 1999 [Wolfgang and Prior, in Freitag (ed.), op. cit., pp. 233–255].

In the past decade, monolithic columns have gained popularity for analytical applications. Instead of using discrete packing particles, a whole polymer block is used as a column. Their continuous homogeneous structure provides fast mass-transfer rates and very high flow rates inside the column [Strancar et al., in Freitag (ed.), op. cit., pp. 50–85]. Thin monolithic disks with affinity binding are capable of fast chromatographic separations. They act much as affinity membrane chromatography cartridges do. To utilize long monolithic columns for process-scale separations, a breakthrough in column fabrication is needed to produce large columns suitable for commercial applications.

Sequencing of Chromatography Steps The sequence of chromatographic steps used in a protein purification train should be designed such that the more robust techniques are used first, to obtain some volume reduction (concentration effect) and to remove major impurities that might foul subsequent units; these robust units should have high chemical and physical resistance to enable efficient regeneration and cleaning, and they should be of low material cost. These steps should be followed by the more sensitive and selective operations, sequenced such that buffer changes and concentration steps between applications to chromatographic columns are avoided. Frequently, ion-exchange chromatography is used as the first step. The elution peaks from such columns can be applied directly to hydrophobic interaction chromatographic columns or to a gel filtration unit, without the need for desalting of the solution between applications. These columns can also be used as desalting operations, and the buffers used to elute the columns can be selected to permit direct application of the eluted peaks to the next chromatographic step.

Factors to be considered in making the selection of chromatography processing steps are cost, sample volume, protein concentration and sample viscosity, degree of purity of protein product, presence of nucleic acids, pyrogens, and proteolytic enzymes. Ease with which different types of adsorbents can be washed free from adsorbed contaminants and denatured proteins must also be considered.

Scale-up of Liquid Chromatography The chromatography columns in downstream processing typically are operated in the nonlinear region due to a concentrated or overloaded feed. Their scale-up remains a challenging task. There are two general approaches: (1) the rule-based method using equations for column resizing (Ladisch, op. cit., pp. 299–448) and (2) the computer simulation method using rate models (Gu, *Mathematical Modeling and Scale-Up of Liquid Chromatography*, Springer-Verlag, Berlin-New York, 1995) with simulation software such as Chromulator® to predict column performance for a particular column setting and operating conditions.

PRODUCT POLISHING AND FORMULATION

The product from the final purification unit operation is typically in a liquid fraction containing water, a solvent, or a buffer. Based on the requirement for the final product, they may need to be removed. A solid protein is usually far more stable with a much longer shelf life. Product formulation may also require an excipient to be added. Thus, additional unit operations are needed after the final purification step.

Lyophilization and Drying After the last high-performance purification steps it is usually necessary to prepare the finished product for special applications. For instance, final enzyme products are often required in the form of a dry powder to provide for stability and ease of handling, while pharmaceutical preparations also require high purity, stability during formulation, absence of microbial load, and extended shelf life. This product formulation step may involve drying of the final products by freeze drying, spray drying, fluidized-bed drying, or crystallization (Golker, in Stephanopoulos, op. cit., pp. 695–714). Crystallization can also serve as an economical purification step [Lee and Kim, in Hatti-Kaul and Mattiasson (ed.), op. cit., pp. 277–320] in addition to its role as a unit operation for product polishing.

Freeze drying, or lyophilization, is normally reserved for temperature-sensitive materials such as vaccines, enzymes, microorganisms, and therapeutic proteins, as it can account for a significant portion of total production cost. This process is characterized by three distinct steps, beginning with freezing of the product solution, followed by water removal by sublimation in a primary drying step, and ending with secondary drying by heating to remove residual moisture.

Freezing is carried out on cooled plates in trays or with the product distributed as small particles on a drum cooler; by dropping the product solution in liquid nitrogen or some other cooling liquid; by cospraying with liquid CO_2 or liquid nitrogen; or by freezing with circulating cold air. The properties of the freeze-dried product, such as texture and ease of rehydration, depend on the size and shape of the ice crystals formed, which in turn depend on the degree of undercooling. It is customary to cool below the lowest equilibrium eutectic temperature of the mixture, although many multicomponent mixtures do not exhibit eutectic points. Freezing should be rapid to avoid effects from local concentration gradients. Removal of water from solution by the formation of ice crystals leads to changes in salt concentration and pH, as well as enhanced concentration of the product, in the remaining solution; this in turn can enhance reaction rates, and even reaction order can change, resulting in cold denaturation of the product. If the feed contains a solvent and an acid, the solvent tends to sublimate faster than the acid, causing acidic damage to the protein. With a high initial protein concentration the freeze concentration factor and the amount of ice formed will be reduced, resulting in greater product stability. For aseptic processing, direct freezing in the freeze-drying plant ensures easier loading of the solution after filtration than if it is transferred separately from remote freezers.

In the primary drying step, heat of sublimation is supplied by contact, conduction, or radiation to the sublimation front. It is important to avoid partial melting of the ice layer. Many pharmaceutical preparations dried in ampoules are placed on heated shelves. The drying time depends on the quality of ice crystals, indicating the importance of controlling the freezing process; smaller crystals offer higher interfacial areas for heat and mass transfer, but larger crystals provide pores for diffusion of vapor away from the sublimation front.

A high percentage of water remains after the sublimation process, present as adsorbed water, water of hydration, or dissolved in the dry amorphous solid; this is difficult to remove. Usually, shelf temperature is increased to 25 to 40°C and chamber pressure is lowered as far as possible. This still does not result in complete drying, however, which can be achieved only by using even higher temperatures, at which point thermally induced product degradation can occur.

Excipients can be used to improve stability and prevent deterioration and inactivation of biomolecules through structural changes such as dissociation from multimeric states into subunits, decrease in α-helical content accompanied by an increase in β-sheet structure, or complete unfolding of helical structure. These are added prior to the freeze-drying process. Examples of these protective agents include sugars, sugar derivatives, and various amino acids, as well as polymers such as dextran, polyvinyl pyrrolidone, hydroxyethyl starch, and polyethylene glycol. Some excipients, the lyoprotectants, provide protection during freezing, drying, and storage, while others, the cryoprotectants, offer protection only during the freezing process. Spray drying can use up to 50 percent less energy than freeze-drying operations and finds application in the production of enzymes used as industrial catalysts, as additives for washing detergents, and as the last step in the production of single-cell protein. The product is usually fed to the dryer as a solution, a suspension, or a free-flowing wet substance.

Spray drying is an adiabatic process, the energy being provided by hot gas (usually hot air) at temperatures between 120 and 400°C. Product stability is ensured by a very short drying time in the spray-drying equipment, typically in the subsecond to second range, which limits exposure to the elevated temperatures in the dryer. Protection can be offered by addition of additives (e.g., galactomannan, polyvinyl pyrrolidone, methyl cellulose, cellulose). The spray-drying process requires dispersion of the feed as small droplets to provide a large heat- and mass-transfer area. The dispersion of liquid is attained by using rotating disks, different types of nozzles, or ultrasound, and is affected by interfacial tension, density, and dynamic viscosity of the feed solution, as well as the temperature and relative velocities of the liquid and air in the mixing zone. Rotating-disk atomizers operate at 4000 to 50,000 rpm to generate the centrifugal forces needed for dispersion of the liquid phase; typical droplet sizes of 25 to 950 μm are obtained. These atomizers are especially suitable for dispersing suspensions that would tend to clog nozzles. For processing under aseptic conditions, the spray drier must be connected to a filling line that allows aseptic handling of the product.

Thickeners and binders such as acacia, agar, starch, sodium alginate, gelatin, methyl cellulose, bentonite, and silica are used to improve product stability and enhance the convenience of the administration of a liquid formulation. Surface-active agents, colors, flavors; and preservatives may also be used in the final formulation (Garcia et al., *Bioseparation Process Science*, Blackwell Science, Malden, Mass., 1999, p. 374).

INTEGRATION OF UNIT OPERATIONS IN DOWNSTREAM PROCESSING

Generally speaking, a typical downstream process consisting of four stages: removal of isolubles, isolation, purification, and polishing (Belter et al., op. cit., p. 5). Cell disruption is required for intracellular products. One or two, and sometimes more chromatography steps will serve as the center-stage unit operations. The steps before them serve the purposes of feed volume reduction and removal of the majority of impurities. The steps after them are polishing and formulation operations. Based on these general outlines, a few rules of thumb may be used (Harrison et al., op. cit., p. 322; Garcia et al., op. cit., p. 358): (1) Reduce the feed volume early in the process, (2) remove the most abundant impurities or the easy-to-remove impurities first, (3) reduce the amount of impurities as much as possible before the delicate high-resolution chromatography steps, and (4) sequence the unit operations that exploit different separation mechanisms.

The purification of proteins to be used for therapeutic purposes presents more than just the technical problems associated with the separation process. Owing to the complex nature and intricate three-dimensional structure, the routine determination of protein structure as a quality control tool, particularly in its final medium for use, is not well established. In addition, the complex nature of the human immune system allows for even minor quantities of impurities and contaminants to be biologically active. Thus, regulation of biologics production has resulted in the concept of the process defining the product since even small and inadvertent changes in the process may affect the safety and efficacy of the product. Indeed, it is generally acknowledged that even trace amounts of contaminants introduced from other processes, or contaminants resulting from improper equipment cleaning, can compromise the product. From a regulatory perspective, then, operations should be chosen for more than just efficiency. The consistency of the unit operation, particularly in the face of potentially variable feed from the culture/fermentation process, is the cornerstone of the process definition. Operations that lack robustness or are subject to significant variation should not be considered. Another aspect of process definition is the ability to quantify the operation's performance. Finally, the ease with which the equipment can be cleaned in a verifiable manner should play a role in unit operation selection. Obviously, certain unit operations are favored over others because they are easier to validate. Process validation was covered in a book edited by Subramanian (op. cit., vol. 2, pp. 379–460).

Keep in mind that some unit operations are not as scalable as others. The evolution of a bench-scale process to production scale will see changes in the types and number of unit operations selected. To configure an effective and efficient bioseparations process, a thorough understanding of the various unit operations in downstream processing described above is a prerequisite. Different process-scale and purity requirements can necessitate changes and may result in quite different configurations for the same product. Existing process examples and past experiences help greatly. Due to regulatory restrictions, once a process is approved for a biopharmaceutical, any change in unit operations requires time-consuming and costly new regulatory approval. This means that an optimized process design is much desired before seeking approval. The involvement of biochemical engineers early in the design process is highly recommended.

INTEGRATION OF UPSTREAM AND DOWNSTREAM OPERATIONS

Upstream fermentation (or cell culture) has a direct impact on the design and optimization of its downstream process. Different media or different operating conditions in fermentation result in a different feed for the downstream process. In the development of new products, optimization of the fermentation medium for titer only often ignores the consequences of the medium properties on subsequent downstream processing steps such as filtration and chromatography. It is imperative, therefore, that there be effective communication and understanding between workers on the upstream and downstream phases of the product development if rational tradeoffs are to be made to ensure overall optimality of the process. One example is to make the conscious decision, in collaboration with those responsible for the downstream operations, of whether to produce a protein in an unfolded form or in its native folded form. The purification of the aggregated unfolded proteins is simpler than that of the native protein, but the refolding process itself to obtain the product in its final form may lack scalability or certainty in its method development.

In some instances, careful consideration of the conditions used in the fermentation process, or manipulation of the genetic makeup of the host, can simplify and even eliminate some unit operations in the downstream processing sequence [Kelley and Hatton, *Bioseparation,* **1**, 303–349 (1991)]. Some of the advances made in this area are the engineering of strains of *E. coli* to allow the inducible expression of lytic enzymes capable of disrupting the wall from within for the release of intracellular protein products, the use of secretion vectors for the expression of proteins in bacterial production systems. Fusion proteins can be genetically engineered to attach an extra peptide or protein that can bind with an affinity chromatography medium [Whitmarsh and Hornby, in Street (ed.), op. cit., pp. 163–177]. This can enhance the purification of an otherwise difficult to purify protein greatly. The cell culture medium can be selected to avoid components that can hinder subsequent purification procedures. Integration of the fermentation and initial separation/purification steps in a single operation can also lead to enhanced productivity, particularly when the product can be removed as it is formed to prevent its proteolytic destruction by the proteases which are frequently the by-product of fermentation processes. The introduction of a solvent directly to the fermentation medium (e.g., phase-forming polymers), the continuous removal of products by using ultrafiltration membranes, and the use of continuous fluidized-bed operations are examples of this integration.

Process economics for biological products was discussed by Harrison et al. (op. cit., pp. 334–369) and Datar and Rosen [in Asenjo (ed.), *Separation Processes in Biotechnology*, Dekker, New York, 1990, pp. 741–793] at length, and also by Ladisch (op. cit., pp. 401–430). They provided some illustrative examples with cost analyses. Bioprocess design software can also prove helpful in the overall design process (Harrison et al., op. cit., pp. 343–369).

Made in the USA
Monee, IL
05 November 2026

67811382R00048